John Watson

Nature and Woodcraft

John Watson

Nature and Woodcraft

ISBN/EAN: 9783337371647

Printed in Europe, USA, Canada, Australia, Japan

Cover: Foto ©berggeist007 / pixelio.de

More available books at **www.hansebooks.com**

NATURE AND WOODCRAFT.

BY

JOHN WATSON, F.L.S.,

AUTHOR OF

"A YEAR IN THE FIELDS," "SYLVAN FOLK," "BRITISH SPORTING FISHES,"
"THE CONFESSIONS OF A POACHER," ETC.

WITH ILLUSTRATIONS BY G. LODGE

LONDON :

A. D. INNES AND CO.,

31 & 32, BEDFORD STREET, STRAND, W.C.

1892.

CONTENTS.

NATURE AND WOODCRAFT.

CHAPTER I.

BIRDS OF PREY.

CUMBRIA is not the primitive spot it once was. As tourists have invaded it, the Eagles and larger birds of prey have left their haunts. The spots which knew the wild white cattle, bears, wolves, and beavers, know them no more; and, by the working of a great natural law, these have become extinct. But if the invasions of a utilitarian age have rid us of the Eagles, they occasionally pay us passing visits in their majestic flights. Among birds the raptors are as kings and princes among men; they hold sway over a wide area, and suffer no intrusion—the raptors, with their

B

clean-cut figures, their bold dash, and glorious
eyes !

The Lake hills long offered an asylum not
only to Eagles, but to all the larger birds of
prey ; and these commonly built among them.
Wordsworth and Wilson mention the Golden
Eagle as breeding in the Lake District; and
in their journals, Gray the poet, and Davy
speak — the one of seeing an eyrie robbed,
the other of watching the birds themselves.
De Quincey has also a note of personal observation.   Raven Crag, the high hills above
Keswick, Thirlmere, and Borrowdale, are sites
of former eyries.   It is asserted by a shepherd
of the district that these Eagles, during the
breeding season, destroyed a lamb daily, to say
nothing of the carnage made on hares, partridges, pheasants, grouse, and the waterfowl
that inhabit the lakes.

At the places above mentioned, the farmers
and dalesmen were careful to plunder the
eyries, but not without considerable risk of
life or limb to the assailant.   In one case,
a man was lowered from the summit of pre-
cipitous rocks by a rope of fifty fathoms, and
was compelled to defend himself from the

attacks of the birds during his descent. Gray graphically describes how the nests were annually plundered, upon one of which occasions he was present. The two species which bred in the district were the White-tailed or Sea Eagle and the Golden Eagle. Wordsworth tells how they built in one of the precipices overlooking Red Tarn, in a recess of Helvellyn, and that the birds used to wheel and hover over his head as he fished these lonely waters. When we last visited the spot, the silence was only broken by the hoarse croaking of a couple of Ravens, the sole remaining relics of the original " Red Tarn Club."

An instance is related of an Eagle which, having pounced on a shepherd's dog, carried it to a considerable height; but the weight and action of the animal effected a partial liberation, and left part of its flesh in the eagle's beak. The dog was not killed by the fall. It recovered from the wound, but was so intimidated that it would never go that way again. The son of the owner of the dog shot, near Legberthwaite, at one of the eagles, which he wounded. This bird was found by a farmer, about a week afterwards, in a state of great exhaustion, the lower mandible having been split, and the tongue wedged between the interstices. The bird was captured and kept in confinement, but became so violent that ultimately it had to be destroyed.

On the Eagles being frequently robbed of their young in Greenup, they removed to the opposite side of the crag. At this place they built for two years, but left it for Raven Crag, within the Coom, where, after staying a year only, they returned to their ancient seat in Eagle Crag. Here they bred annually during their stay in Borrowdale. On the loss of its mate, the other eagle left the district, but returned in the following spring with a fresh one. This pair built during fourteen years, but finally abandoned Borrowdale for Eskdale. Here again they were disturbed, and the female being afterwards shot, the male flew off and returned no more.

Eagle Crag is a grand, towering rock, or collection of perpendicular rocks, connected by horizontal spaces of variously coloured vegetation. Its front is fine, and forms a majestic background to many pleasing scenes. On that part of Eagle Crag which is opposite to

Greenup, the eagles occasionally built their nests. But they were so destructive to the lambs, and consequently injurious to the interests of the shepherds, that their extermination became absolutely necessary. Their building places being inaccessible by climbing, a dangerous experiment was tried. A man was lowered by a rope down the face of the cliff for ninety feet, carrying a piked staff, such as is used by the shepherds, to defend himself against the attack of the birds while he robbed the nest of eggs or eaglets. If birds, their possession was to be his remuneration; if eggs, every neighbouring farmer gave for each egg five shillings. The nest was formed of branches of trees, and lined with coarse grass and bents from the neighbouring rocks. The Eagles sometimes flew off with lambs a month old, and in winter frequented the head of the Derwent, where they preyed upon waterfowl.

The White-tailed Eagles bred upon the rocks of an escarpment overlooking the sea, and fed upon gulls and terns. The vast peat-mosses which stretched away for miles below abounded in hares and grouse, among which the birds made terrible havoc. Year after year they carried off their young from the same cliffs, but now return only at rare intervals, or when storm-driven.

The Peregrines have the eagles' eyries, and are eagles in miniature. Sea-fowl form their food in summer, as do ducks, plover, and game in winter. At this latter season, the Osprey, or Fish-hawk, comes to the bay and the still mountain tarns, adding wildness to the scenes which its congeners have left never to return.

We are lying on the outskirts of a dark pine-wood interspersed with firs and hollies. A large bird has just flown into that clump of trees on the hillside opposite. There it sits on a dead bough, with its mottled breast towards us, and restless head quickly turning from side to side. Against the dark-green foliage we see the bright orange of its tarsi, and know it to be a Sparrow-hawk. As it flies from the clump, a pair of missel-thrushes and a flock of smaller birds follow in its wake, but dare not mob it. It swoops as one approaches too near, glides upward, and pursues its way, scarcely deigning to note the screeching mob. The hawk glides silently into the wood, threading its sinuous way through the trees, and takes up its position in the centre. The cooing of wood-pigeons seems to excite it, and it makes a circuit, skimming over the ground at the height of a few yards. Then, as something in the grass attracts it, it beats the air with its pointed wings, and, depressing its tail, hangs as if suspended. In a second it falls, just as a lark shoots from a tuft to seek the shelter of a thick thorn-hedge. The hawk follows, and beats the bush, first on one side, then the other; but the trembling lark cannot be frightened out of its stronghold, and the bird, finding itself baffled, skims along as before.

Round and round the wood it flaps, now sweeping low over the trees, anon hanging motionless. A number of chaffinches are picking among the corn unconscious of the presence of an enemy. Suddenly the hawk darts round the corner of the wood into the midst of the terrified flock, clutches one in its talons, and is off straight and swift across the country, staining with a deeper scarlet the ruffled plumage of its captive.

Let us in imagination follow this bold spirit of the air to some such plantation as it has

just left, and there, on the topmost branches
of a pine, somewhere near the centre of the
wood, we shall find its nest. It is bulky,
having been repaired annually for years, and
somewhat neatly constructed of fir branches.
It is nearly flat, and on its edge is the
chaffinch, torn limb from limb and cleanly
plucked. Those four screaming demons clothed
in down are young Sparrow-hawks, and never-
satisfied things they are. We descend the tree,
just keeping in mind a rotten bough, and leave
the young ones to enjoy their feast. Yonder
on an ash-stump sits the female, quietly watching our movements, to return when we are
gone.

The spot on which we lie is a haunt of the
Kestrel—a perpendicular limestone escarpment
which rushes sheer down fifty feet for a mile
along its front. Below is a flood of green,
patched by the mellower tints of rolling crops.
On one hand mosses and silvery sands stretch
away far beneath us, and on the other rise the
mist-capped peaks of the hills. What a scene
of peace and contentment! White farmhouses
lie like spots of sunlight on the dark green
landscape, each embowered in its clump of

sycamores, which serves to shade and keep
the dairy cool. A limestone road winds its
sinuous way far out among the brown heather,
almost as far as the eye can reach. There the
Greenwash, like liquid silver, flows on until
it is lost in the sands to the south. It sees
as it goes the haunts of gulls, terns, and
herons. Now our attention is attracted by two
small blue pigeons that are flying along the
base of the cliff. After watching for a moment,
we know them to be the beautiful Rock-dove
from which our domestic stock is descended.

We are lying on the turf, when a shadow
floats past us. We look up, and there comes
the pleasant cry, *Kee, kee, keelie.* Suspended
above us and hovering in the wind is the
Kestrel. So quickly do its wings vibrate, that
we can scarce detect the motion as the bird
hangs against the blue. It hovers a while,
then flies to a short distance, and is again
attracted by a stirring in the tangled turf
of grass and bents. Poising itself for a second,
it drops like a stone on closing its wings, which
it just slightly expands again as it takes a
mouse in its talons and flies off to the cliff.
When this morsel has been devoured, the male

and female fly from the nest, and perform—just for the love of exercise, it would seem—a series of aërial evolutions that it would be impossible to describe. The nest in this instance is upon the projecting ledge of a rock midway down the scaur, and protected from sight and the sea-winds by an old, twisted yew.

We are scrambling among the crags in search of Alpine plants, when a large bird of prey advances on the wing. At a distance the under parts appear to be white, but the bird flying directly over at a height of sixty feet enables us to see distinctly the dark bars across the feathers of the abdomen. Its flight is a sort of flapping motion, not unlike that of the Ring-dove; and we can see its head turned rapidly in various directions, the eye at the same time peering into the crannies of the rocks and ghylls, in search of any skulking prey. The Peregrine is marked by dark streaks proceeding from the corners of its bluish-grey back, and by the transverse bars just mentioned. It will dash through a flock of wild ducks or a covey of partridges, wounding several in its *sortie*, but eventually carrying off the one selected with unerring aim.

A noble bird is the Peregrine, with its glorious eyes, wild, restless, and changeful! This bird is the falcon of the royal falconers; its mate, the tiercel. Among all our British birds the Peregrine ranks first; for strength, and courage, and speed it has no compeer. Rooks clamour and arrange themselves in battle array at its approach; other hawks fly off to the covert; small birds of every species seek the thickest shelter, and farm-yard poultry their roost, as it sails in mid-air down dale. Even the eagle suffers itself to be mobbed by the comparatively small Peregrine without offering any retaliation.

We advance over the heather, and there, skimming towards us, is a large bird — a Harrier. The species cannot be doubted, as it flies near the ground, working it as a hound or a setter would do. Now it stoops, glides, ascends, stoops again, and shoots off at right angles. It rounds the shoulder of the hill and drops into a dark patch of ling. A covey of young grouse whirr heavily over the nearest brae, but the Marsh Harrier remains. It has struck down one of the "cheepers" and is dragging its victim to the shelter of a furze

bush. The wonderful evolutions and movements in which the bird indulges, its sudden
swoops, its ascending and descending, seem all
regulated by its tail.

A male and female Harrier generally hunt
together, and afford a pretty sight as they
" harry " the game, driving it from one to the
other, and hawking in most systematic fashion.
They thoroughly quarter the ground previously
marked out, and generally with success. When
they hawk the quiet mountain tarns, their mode
is regulated according to circumstance. In
such instances, they not unfrequently sit and
watch, and capture their prey by suddenly
pouncing upon it.

The great grouse poachers of the moors are
the beautiful little Merlins. They work together over the heather like a brace of well-
broken pointers. Not an object escapes them ;
however closely it may conform to its environment, or however still it may keep, it is detected
by the sharp eye of the Merlin and put away.
The miniature falconry in which this bird
indulges on the open moorlands, where nothing
obstructs the view, is one of the most fascinating sights in nature.

The "red hawk" is plucky beyond its size and strength, and will pull down a partridge, as we have witnessed repeatedly. The young of moorfowl, larks, pipits, and "summer snipe" constitute its food on the fells. It lays four bright red eggs in a depression among the heather, and about this are strewn the remains of its prey. To show to advantage this smallest of British falcons ought to be seen in its haunts. It is little larger than a thrush, and in the days of falconry was flown by ladies at larks, pipits, pigeons, and occasionally partridges. On the moorlands it may be seen suddenly to shoot from a stone, encircle a tract of heather, and then return to its perch. A lark passes over its head, and its wings are raised and its neck outstretched ; but it closes them as if unwilling to pursue the bird. Then it flies, skimming low over the furze, and alights on a granite boulder similar to the one it has just left. As we approach, the male and female flap unconcernedly off, and beneath the block are remains of golden plover, ling-birds, larks, and young grouse.

It was a wise legislative proceeding that granted a double protection to Owls, for of

all birds, from the farmer's standpoint, they
are the most useful. They hunt silently and
in the night, and are nothing short of lynx-
eyed cats with wings. The benefit they confer
upon agriculturists is almost incalculable, as is
susceptible of easy proof.

It is well known that Owls hunt in the
night, but it may be less a matter of common
knowledge that, like diurnal birds of prey,
they disgorge the hard 'indigestible parts of
their food in the form of elongated pellets.
These are found in considerable quantities
about the birds' haunts, and an examination
of them reveals the fact that Owls prey upon
a number of predacious rodents, the destruction of which is directly beneficial to man.
Of course, the evidence gained in this way is
incontestable ; and to show to what extent
Owls assist in preserving the balance of nature,
it may be mentioned that an examination of
seven hundred pellets yielded the remains of
sixteen bats, three rats, two hundred and thirty-
nine mice, ninety-three voles, one thousand
five hundred and ninety shrews, and twenty-
two birds. These remarkable results were
obtained from the common Barn Owl ; the re-

mains of the birds consisting of nineteen sparrows, one greenfinch, and two swifts.

The Tawny and Long-eared Owls of our woodlands are also mighty hunters, and an examination of their pellets shows equally interesting evidence. It must be remembered in this connection that Britain is essentially an agricultural country; and that if its fauna is a diminutive one, it is not the less formidable. We have ten tiny creatures, constituting an army in themselves, that, if not kept under, would quickly devastate our fields. These ten species consist of four Mice, three Voles, and three Shrews—individually so tiny, that any one species could comfortably curl itself up in the divided shell of a chestnut. But farmers well know that if these are small they are by no means to be despised. When the corn crops are cut, and the hay housed, the Field-vole and the Meadow-mouse are deprived of their summer shelter. Of this the Barn Owl is perfectly aware, and at evening may be seen sweeping low over the meadows, seeking what it may capture—with what results we have already seen.

Much unnatural history has been written of

Owls, and unfortunately most people take their ideas of them from the poets. It is unnatural to assert, as they do, that Barn Owls ever mope, or mourn, or are melancholy. Neither are they grave monks, nor anchorites, nor pillared saints. A boding bird or a dolorous! Nonsense! they are none of these. They issue forth as very devils, and, like another spirit of the night, sail about seeking whom they may devour.

Poets write by day, and Owls fly by night; and, doubtless, Mr. Gray and his school have their opinion of owls from staring at stuffed specimens in glass cases, or at the living birds surprised in the full light of day, when they will be seen blinking, nodding, and hissing at each other, very unlike the wise representatives of Minerva. Christopher North is the only writer who has done justice to Owls—or justice to poets, for the matter of that—by his denunciation of their epithets and false images. He knew well that the White Owl never mopes, but holds its revels through the livelong night, when all else is hushed and still. Most birds are stoics compared to Owls, and those who cultivate their acquaintance know that they have no time wherein to make their poetical complaints to the moon. Poets should not meddle with Owls. Shakespeare and Words-worth alone understood them; by all else they have been scandalously libelled—from Virgil to the Poet Close.

The Barn Owl, when she has young, brings to her nest a mouse about every twelve minutes; and, as she is actively employed both at evening and dawn, and as male and female hunt, forty mice a day is the lowest computa-tation we can make. How soft is the plumage of the Owl, and how noiseless her flight! Watch her as she floats past the ivy tod, down by the ricks, and silently over the old wood; then away over the meadows, through the open door, and out of the loop-hole of the barn; round the lichened tower, and along the course of the brook. Presently she returns to her four downy young, with a mouse in one claw and a vole in the other, soon to be ripped up, torn, and eaten by the greedy, snapping imps. Young and eggs are not unfrequently found in the same nest.

If you would see the midday *siesta* of these birds, climb up into some hay-mow. There, in an angle of the beam, you will see their

c

owlships, snoring and blinking wide their great round eyes. Their duet is the most unearthly, ridiculous, grave noise conceivable; unlike anything you ever heard. There they will stay all day, digesting the mice with which they have gorged themselves, until twilight, when they again issue forth upon their madcap revels. This clever mouser, then, has a strong claim to our protection ; so let not idle superstition further its destruction.

CHAPTER II.

I WELL remember what, as a country lad, impressed me most upon my first visit to London. It was the recollection of the fact that I had, during the small hours of the morning, stood alone in the Strand. I had walked into the city from a suburban house. As I paced rapidly along the pavement my footsteps echoed, and I listened to them until, startled, I came to a dead stop. The great artery of life was still; the pulse of the city had ceased to beat. Not a moving object was visible. Although I had been bred among the lonely hills, I felt for the first time that this was to be alone; that this was solitude. I felt a sense such as Macaulay's New Zealander may experience when he sits upon the ruins of London Bridge; and then for the first time I knew whence the inspiration, and felt the full force and realism of Wordsworth's exclamation, " O God! the very houses seem asleep." Than this I could detect no definite sound—only that vague and distant hum which for ever haunts and hangs over a great city.

Such a time of quiet can never be observed in the country.   It matters not as to time or season; there seems no absolute and general period of repose.   There is always something abroad, some creature of the field or wood, which by its voice or movements is betrayed. And, just as in an old, rambling house, there are always strange noises that cannot be accounted for, so in the by-paths of nature there are innumerable sounds which can never be localized.   To those, however, who pursue night avocations in the country — gamekeepers, poachers, etc.—there are always calls and cries which bespeak life as animate under the night as that of the day.   This is attributable to various animals and birds, to beetles and night-flying insects, and even to fish. Let us track some of these sounds to their source.

We are by the covert-side, and a strange churring comes from the glades. Waiting silently beneath the bushes, it approaches nearer and nearer, until a loud flapping is heard among the nut-bush tops. The object approaches quite closely, and we can see that the noise is produced by a large bird striking its wings together as they meet behind. Even in the dark we detect that each wing is crossed by a definite white bar. Had we the bird in our hand, we should see that it seemed a connecting link between the Owls and the Swallows, having the soft plumage and noiseless flight of the one, and the wide mouth of the other. The noise it produces among the trees is probably to disturb from off the bushes the large winged moths upon which it feeds. This is the Nightjar or Goat-sucker. The latter name it has from a superstitious notion that it sucks goats and cows, founded probably upon the fact of its wide gape. It is certain that these birds are often seen flitting about the bellies of cattle as they stand knee-deep in summer pastures. The reason of this is obvious, as there insect food is always abundant.

Coming from out the woods the short sharp

bark of a Fox is heard, and this is answered
at intervals by the vixen. Rabbits rush across
our path, or rustle through the dead leaves,
their white scuts showing as vanishing-points
in the darkness. The many-tongued Sedge-
bird which tells her tale to all the reeds by
day, prolongs it under the night. Singing
ceaselessly from the bushes, she chatters garru-
lously or imitates the songs of other birds;
until my old angler friends call her the " fisher-
man's nightingale." When by the covert-side,
one of the calls which one constantly hears is
the crowing of cock pheasants; this is indulged
in the densest darkness, as is sometimes the
soft cooing of the wood-pigeons.

Both pheasants and cushats sleep on the
low lateral branches of tall trees, and from
beneath these the poacher often shoots them.
He comes when there is some moon, and with
a short-barrelled gun and a half-charge of
powder drops the birds dead from below. One
of the greatest night helps to the gamekeeper
in staying the depredations of poachers is the
lapwing. This bird is one of the lightest
sleepers of the fields, starting up from the
fallows and screaming upon the slightest alarm.

Poachers dread the detection of this bird, and the keeper closely follows its cry. A hare rushing past will put the plover away from its roost, and when hares act thus there is generally some good cause for it.

At night the waterside is productive of life, and here it is most varied. Turning a bend of the stream, a sentinel Heron that has been standing watchful on one leg, rises, and flaps languidly away down the river reach. The consumptive figure of the gaunt bird stands by the stream through all weathers. He knows not times nor seasons, and is a great poacher. In the wind, when taking his lone stand, his loose fluttering feathers look like driftstuff caught in the bushes. He reminds one of the consumptive, but, unlike him, has wonderful powers of digestion, and withal an immense capacity for fish. Woe to the luckless mort or trout that comes within reach of his formidable pike, or to the attacking peregrine that he attempts to impale on his bill. The Heron is essentially a wanderer, and, like Wordsworth's immortal leech-gatherer, he roams " from pond to pond, from moor to moor."

Passing the remains of an old Baronial Hall,

the piercing screech of a Barn Owl comes from
a dismantled tower. Here the white owls have
lived time out of mind, and we have seen
and heard them, asleep and awake, through
every hour of the day and night. The Barn
Owl is the " screech " owl of bird literature, the
Brown Owl the true hooting owl—the former,
however, hoots at times. This species is found
in old and heavily timbered districts, and particularly loves the dark and sombre gloom of
resinous pine-woods.

One of the most piteous sounds that is borne
on the night is the hare's scream when it finds
itself in the poacher's nets. It resembles nothing
so nearly as the cry of a child, and when it suddenly ceases, we know that the wire snare has
tightened round its throat. All night long
Crake answers Crake from the meadows, appearing now at our feet, now far out yonder.
Like the Cuckoo, the Cornrail is a bird oftener
heard than seen ; it is of hiding habits, and
finds a secure and snug retreat in lush summer
grass. Beneath the oaks, bats encircle after
night-flying insects, and there by the stream-
side are clouds of gauzy ephemerœ. The wild
whistle of a Curlew comes from high over-

head, as the bird flies through the night to
its far-off feeding-ground. In the fall of the
year, multitudes of migratory birds pass over;
we "hear the beat of their pinions fleet,"
but their forms we cannot see. If only,
however, we hear the cry of their voices,
falling dreamily through the sky, the species
is easy of identification. If we approach the
reed-beds silently, we may hear the hoarse
croak of the frogs; or springing wild ducks,
as they beat the air with their strong wings.
Emerging from the waterside to a belt of
coppice, we are again reminded how lightly
the creatures of the fields and woods sleep.
The faintest rustle brings chirping from the
bushes, and in the densest darkness even some
of the delicate wood-birds sing—not only the
Sedge and Grasshopper Warblers, but from
the willows come the lute-like mellowness and
wild sweetness of the Blackcap, another night
singer.

There are some conspicuously white flowers
which only emit their fragrance at night, and
these have their own particular night-flying
insects to fertilize them. From among the
ground-weeds rises at intervals the sweet

smell of the purple Dame's-violet, the Evening
Campion, and bifoliate Orchis.

Besides those enumerated, many other sounds
there are, known only to dwellers in the country,
or those who have brushed the beads from the
long grass during the short summer nights.

CHAPTER III.

BRITISH WILD PIGEONS.

" WILD pigeon " is a term loosely applied. In reality there are five British species. These comprise the Ring-dove, the Stock-dove, the Rock-dove, the Turtle-dove, and the Passenger Pigeon. The beautifully burnished Ring-dove is the wood-pigeon, or cushat—a large, handsome bird, abundant everywhere throughout the country, and often seen in immense flocks in the autumnal months. The Stock-dove is a smaller bird. The Rock-dove is a little blue pigeon, from which it is generally believed that all our domestic varieties are descended. The Turtle-dove is a summer migrant to our shores; and the American Passenger Pigeon can only be looked upon as a casual visitant of very rare occurrence.

To dwellers in the country the Ring-dove is one of the best-known wild birds, and is yearly becoming more common. Flocks amounting to hundreds may be seen flying together, ranging over wide tracts of country in search of food. When thus banded, they feed upon acorns, beech-nuts, grain, and the leaves of green crops.

There is no question as to the devastation which one of these flocks brings with it. It is said that 1020 grains of corn have been found in the crop of a single bird ; and that in East Lothian, where less than a century ago the species was unknown, 130,440 birds have been killed within seven years, and 29,000 in twelve months, without apparently decreasing the numbers. I myself shot one of a flock of Ring-doves, the crop of which contained 67 acorns, while in that of another bird of the same species, 80 beans were found. Lord Haddington examined the crops of four Ring-doves with the following results : That of the first contained 144 field-peas and seven large beans ; the second, 231 beech-nuts ; the third, 813 grains of barley ; the fourth, 874 grains of oats and 55 of barley. From the above, the damage done

by these birds may be guessed at, especially
when it is remembered that they feed three
times daily. The causes which contribute to
swell the number of wild pigeons in this country
are well known. More green crops are now
grown for winter use than formerly; and
thus food is provided during a period of the
year when in times past it was unattainable.
Then the economy of fir or spruce plantations
is better understood, and these afford precisely the harbour which the wood-pigeon loves
best.

For eight months of the year the Ring-dove
is gregarious. After the last young are
hatched, the birds begin to flock, as the farmer
discovers, about the time of the ripening of
the corn. Often the grain stooks swarm with
them, and they rise in clouds as they are
driven from the fields. Early in autumn numbers are killed. They soon become shy, and
then few birds are so wary or so difficult of
approach. After the garnering of the grain,
pigeons have another great harvest in beech-
mast and acorns. For the most part the
" pannage " is picked from the ground, but
sometimes the birds fly up into the trees and

take their spoil from the branches. The mast of oak and beech constitutes the staple food for many months, and at this time they become extremely fat. Long after the snows and frost have set in, flocks of Ring-doves may be seen scratching among the fallen leaves in search of nuts. When this supply begins to fail, the pigeons resort to turnip-fields, where they not only feed upon the green " tops," but upon the bulbs themselves. The Ring-dove is essentially a soft-billed bird, and is therefore unable to break through the outer rind of the root ; but the way is too frequently laid open by rabbits, hares, and rooks.

The immense flocks of wild pigeons which are now so common in winter are not all British-bred birds : in autumn large numbers of them arrive from over the North Sea to winter upon our shores. With returning spring and sprouting woods, the flocks begin to break up. They are now less frequent in the lowlands, and go up the sides of those valleys which are planted with fir-woods and coppice. For resinous woods they seem to have a special liking, and soon from out the pines may be heard the coo-coo-cōō as one of the pairing notes. At this time the birds feed almost exclusively upon green food, devouring tender shoots and leaves in enormous quantities. They like green or sprouting corn, and the leaves of various varieties of field-clover; numbers of birds have been shot in early summer with their crops distended with gooseberries.

The Ring-dove has two or three broods during the season, and nesting operations commence early. In open seasons the nest is often completed by the end of March. The birds seem to shun the light, and the nest is oftenest found in the more gloomy parts of the wood. This is placed low upon a bough or between a bough and the trunk; and so wicker-like is the platform of fir-twigs, that through the frail structure the two white eggs may often be seen from beneath. It is a curious fact that not unfrequently the Sparrow-hawk and Wood-pigeon build within a few yards of each other, and apparently live at peace. Both parents take their share in sitting, and incubation lasts from sixteen to eighteen days. Very little time is lost between the first and second broods, and eggs and young are occasionally found in the same nest.

On account of its frequenting similar nesting-ing-places, the Stock-dove was at one time con-founded with the Rock-dove, which, except in this particular, it hardly resembles. Although the former often associates and feeds with the Ring-dove, it can easily be told from that bird by the absence of white about the neck. As a species, the Stock-dove is rapidly extending its range in a northerly and westerly direction. This has been particularly observed of late in Western Europe, whilst in our own country it is most marked. .

Like most pigeons, the Stock-dove rapidly adapts itself to circumstances, breeding and haunting very dissimilar spots in different localities. In the south, it betakes itself to the " stocks " of pollarded trees; and it is probably this circumstance to which it owes its name. In the north, the building of this species in stocks or hollow trees hardly holds good. It is fond of resorting to rabbit burrows; and sometimes sheld-ducks, stock-doves, and rabbits may be found forming one community and breeding in the closest proximity. In autumn, Stock-doves not only fly in flocks, like their congeners, but associate with large bodies of

Ring-doves, scouring the country in search of food. In the air it is difficult to distinguish between the two species; but, if closely observed, the Stock-dove is seen to be the smaller bird, and to fly with greater lightness and speed than the Ring-dove. The birds alight in the same feeding-grounds, though experiments have proved that they do not invariably feed upon the same food. For instance, in the case of two birds killed at one shot, the crop of a Ring-dove contained clover-leaves, turnip-tops, and pulp; whilst that of the Stock-dove was without a leaf of clover, but contained a quantity of charlock, barley, and the seeds of weeds.

From this fact it may be seen that wild pigeons may do good as well as harm. Like the Ring-dove, the Stock-dove is an early breeder. The pairing notes are not so pleasant, being neither so round nor so full as the cooing of the cushat. When the bird breeds in a burrow, only a few sticks or bents are collected, and sometimes the eggs are simply deposited on the dry soil. These are creamy white; they hatch in about seventeen days; and the young quickly come to maturity. No sooner do these quit the nest than another

D

couple of eggs is laid; and this goes on until
the end of July. The nests are from two to
three feet down the burrow; and in flying
homewards the birds drop right at the mouth
of the hole, and not at a distance, as is sometimes stated. There is abundant evidence that
the food of this species consists less of cultivated crops and more of the seeds of weeds and
grasses, beech-mast, and acorns. If this characteristic is continued as the species increases in
number (and this it is certainly doing), the
Stock-dove will become as great a friend to
the farmer as its congeners are his enemies.
All weed-devourers are beneficial to agriculture.

The Rock-dove is a beautiful blue pigeon,
and the smallest of the five British species.
As its name implies, it builds in rocks, and for
the most part among the cliffs of the sea-coast.
It is often confounded with its congeners, on
account of their occasionally nesting in like
situations; but it may easily be distinguished
from the Ring-dove by the absence of the ring
round the neck, and from the Stock-dove by its
white bodice. When the bird builds in escarp-
ments and can be observed from above, this
last characteristic and the pale-blue of its

feathers render it very distinctive as it flies
along the face of the cliff. The Rock-dove is
rare upon the east coasts, and but seldom
strays to the south. It is along the north-
western seaboards of Britain that it is found,
the coast here being broken and indented and
abounding in caves. Here the birds breed in
vast numbers, surging out when alarmed in
such flocks as to make a sort of subdued roar
with their wings. Enter one of these caves in
the breeding-season, and you will see the birds
covering every shelf and cranny, their white
eggs showing beneath as the body is slightly
raised at the first sign of danger.

Like its congeners, the Rock-dove breeds
early in the season, from the first days of April
right on through summer. Two small, pure-
white eggs are laid; and in whatever situation
the nest may be, it is composed of roots, dried
grass, and seaweed. This species, like the
Ring-dove and Stock-dove, is a grain and seed
feeder, and, like them, has an enormous capacity
for food. In search of it the Rock-dove makes
short migrations, though it never appears in
great flocks. It is partial to wild land, especially if it be near the coast. Saline water it

delights in, and on salt-marshes it may frequently be seen feeding. Here it devours
the lower forms of animal life, especially tiny
shelled-snails. It flies and feeds with the other
wild pigeons, though its food is far more various
than theirs ; it also eats the roots of couch-
grass, and the seeds of several troublesome
weeds. Its voice is different from that of the
other wild pigeons; and like the various domestic breeds it has a decided aversion to settle
on trees.

The delicate Turtle-dove is essentially a bird
of southern haunts, and only comes to Britain
as a summer migrant. In the woods of the
south it is not uncommon ; though, like the
Nightingale and some other birds, it seems
gradually to be extending its range. The
Turtle-dove arrives on our shores with the
wood-warblers about the beginning of May,
and if the weather is seasonable its soft notes
are soon heard in the copses. It delights in
open woods with sunny glades, and it is from
the darker-foliaged trees in such situations that
its subdued cooing may first be heard. If the
birds be approached without alarming them,
the interesting evolutions in which they at this

time indulge may be observed. The female dove sits passively upon a bough of pine or fir while the male describes a series of circles round her. At first he coos softly; but, after a preparatory pouting to his mate, he puffs out his throat-feathers and indulges in what is very much like a hearty laugh. This continues while the wicker-like nest is building, and even after the two pinky-white eggs are laid. Incubation lasts rather over a fortnight, and it is rare in this country that the Turtle-dove rears more than one brood. It is by far the most delicate of the pigeons, and is peculiarly susceptible to cold. It comes when spring is well advanced, and at the first suggestion of autumn cold it flocks and leaves our shores. Family parties may be seen by the middle of September, and by the end of that month few remain.

Although the Turtle-dove eats grain, it devours enormous quantities of the seeds of weeds, and also those of many objectionable grasses. It is partial to vetches, rape, charlock, wild mustard, and especially haunts the pulse-fields. It is often seen busily employed upon the fallows, and, like the rest of the wild pigeons, frequents fields containing root-crops, especially

in hot weather. Here it finds both food and
shelter, and to such spots small family parties
go before flocking in early autumn. Like most
birds, the Turtle-dove migrates in the night.
Although delicately formed, it is possessed of
considerable powers of flight, and is by no
means easy to shoot. In the woods it winds
its sinuous way through the branches in a
remarkably quick manner, though when clear
of cover it flies strongly and swiftly.

Care ought to be exercised in sifting supposed instances of the Turtle-dove's breeding in
Northern Britain. All through the country the
bird is commonly kept in confinement, and in
summer escaped birds sometimes manage to rear
their young in the open. A case came under my
notice where a pair of tame birds not only built,
but reared *two* broods in a shrubbery. Seeing
the success of this accidental experiment, six
additional pairs were allowed their liberty in
the following spring, each of which bred in the
evergreens of the garden and were fed daily
on the lawn. These birds, however, obtained
much natural food, and by autumn there was
quite a colony of them about the house. The
younger birds showed a wild disposition, though

most were caught and taken indoors for the winter.

The last of the wild pigeons is in some respects the most interesting. This is the American Passenger Pigeon, and, upon what is generally considered sufficient evidence, it is now recognized as a British bird. Several examples have occurred, and whilst some of these were probably " escapes," others doubtless were wild birds. These had perfect plumage, were taken in an exhausted condition, and their crops showed only the slightest traces of food. As is well known, the Passenger Pigeon is a bird of immense power of flight, and in its overland journeys often flies at the rate of a mile a minute. Wild birds, however, can only come from America; and this opens up the interesting question as to the possibility of birds crossing the whole Atlantic without resting. Naturalists of the present day say that this feat is not only probable, but that it is actually accomplished by several wild birds. Mr. Darwin somewhere asserts that one or two of them are annually blown across the ocean; and it is certain that half a dozen species have occurred upon the west coasts of England and Ireland

which are found nowhere but in North America.
Mr. Howard Saunders states that Passenger
Pigeons are often captured in the State of New
York with their crops still filled with undigested
grains of rice that must have been taken in the
distant fields of Georgia and South Carolina ;
apparently proving that they had passed over
the intervening space within a few hours.

It certainly seems remarkable that a bird
should have the power of flying over four
thousand miles of sea ; but recently two different writers have recorded the fact that they
have noticed pigeons settle upon the water to
drink, and then rise from it with apparent ease.
And Mr. Darwin says that, where the banks
of the Nile are perpendicular, whole flocks of
pigeons have been seen to settle on the river and
drink while they floated down stream. He
adds that, seen from a distance, they resembled
flocks of gulls on the surface of the sea.

The Passenger Pigeon is one of the handsomest
of its kind. The accounts of its migrations in
search of food are known to all. It is said to
move in such flocks as to darken the earth as
they pass over, and that one of these columns
brings devastation wherever it comes.

CHAPTER IV.

PROBABLY every one bred in the country will remember some tale or legend attaching to particular trees in the parish, which are looked upon by the villagers with superstitious fear. There is the " Hanging Tree " at the four-road-ends; the " Shrew Ash;" the oak from which the great Quaker preached; and the Red Rowans, with their strange attributes. Some of this legendary lore is more or less local, while some prevails throughout the country. It is natural enough that that tree which enjoys the greatest longevity should have most associations; indeed, few enjoy much notoriety before they begin to show signs of dissolution.

The myth attaching to the " spirit-blasted " Oak of Nassau Park, in North Wales, is a type

of many others. This is an immense tree of
nearly thirty feet in girth, into the hollow bole
of which Owen Glendower threw Howel Sele,
a Welsh chieftain, whom he had wounded in
quarrel. This was while hunting; and Howel,
not making his appearance, search was made
for him, but in vain. The oak kept its secret
until, long after, the truth was told by one
Madoc, an accomplice in the crime, at the suggestion of the dying Glendower, and, upon
search being made, the skeleton of Sele was
found in an upright position, the bones of the
hand grasping a rusted sword.

But it is not alone to the oak that these
mythical stories or superstitions attach, for has
not almost every country parish its " Shrew
Ash?" The velvety Shrew is a perfectly harmless little creature. Nevertheless, our ancestors
looked upon it as an " uncanny beastie," and
invested it with a most malignant character.
Nothing in the fields was safe from its evil influence, and whenever it came near a domestic
animal, the latter immediately lost all power of
limb, and was unable to move. The more
superstitious of our forefathers were quick to
find antidotes for these dark workers, and the

remedies were more often than not of a cruel
nature. In this case a growing ash, usually of
large dimensions, was sought out, and a hole
bored into the trunk. This done, some old hag
would mumble an incantation, and the poor
little Shrew-mouse was thrust in alive, after
which the hole was securely plastered over with
clay. From that time the tree was treated with
great veneration, and became a "Shrew Ash."
For now the parish was armed against the mouse,
and when any creature of the field was afflicted,
it had but to be touched with a twig of such
tree, and it became whole immediately. There
is scarcely a county in England which has not
a number of these trees showing their deep
scars—if, indeed, the practice itself has yet
died out.

For "making butter to come" probably
every pretty milkmaid has a different device.
But the commonest of all is to have inserted
in the side of the churn a bit of wych-hazel,
or wych-elm, though why, nobody seems to
know. The magical properties of this tree, it
has been suggested, are owing to the mere
coincidence between the names "wych" and
"witch." The wood, too, of the Rowan tree, or

Mountain-ash, in the past possessed wonderful magical properties, but would seem to have lost its virtues of late. It was a sovereign charm against the evil machinations of witchcraft, and in the North to-day the peasants keep it about their persons, together with "magpie wings and sic-like things," as these are supposed to "guard puir folk fra' harm." It is also hung over doors and stables to neutralize the spells of witches and warlocks.

Another kind of "myth" connected with trees is frequently due to mistakes of names, or to ignorant tales told to children in their nursery days. Probably every child who has lived within the influence of trees has had pointed out to it the sycamore as the tree into which Zacchæus climbed in his anxiety to see Christ as He passed on his way to Jerusalem. But the Bible "sycamore" is a fig-tree, while ours is a species of maple.

This is the tree whose leaves are covered with a saccharine excretion known as honey-dew, and usually supposed to be a plant production. It is, however, an exudation of a species of aphis which haunts the tree. Another similar "myth" is that the edible and horse-chestnuts are nearly akin, and that they belong to the same genus. Botanically, they could hardly be further apart, for they belong to different natural orders.

Our best Western rendering of Eastern "palms" are the white silky catkins of the Goat-Willow or Saugh; and there is probably not a country child when gathering these on the Sunday next before Easter, but thinks they are the self-same palms which those other children bore before Our Saviour as he walked into Jerusalem. The poisonous properties of the foliage of the Yew are well known, though this fact does not extend to the viscid scarlet berries, as is often supposed, and they may be eaten without the least harm. The Mistletoe growing upon the oak is generally treasured in youthful minds as a school-book fact, but in nature it is of the rarest occurrence. And, unless the plant has strangely altered its habitat since the time of the Druids, it is hardly probable that the priests stripped it from the trees of their sacred groves in any great quantity.

The legend of the "Glastonbury Thorn" is connected with a kind of tree not yet mentioned. It is told how Joseph of Arimathæa, coming to preach the Gospel in Britain, landed on the then Isle of Avelon, attended by twelve like-minded followers. Finding himself weary and heavy with sleep, Joseph thrust his staff into the ground, and lay down. When he awoke, he found that the staff had taken root, had branched, and burst forth into a flood of white blossom. This miracle having been wrought in a single night, was interpreted by the holy men to mean that after their wanderings, their staff and support being taken from them, here they must fix their abode, and fulfil their sacred mission. This they did, building a religious house, which, through the piety of. succeeding generations, increased to its subsequent magnificence.

Legend has also been busy with the Apple-tree and the Aspen. The fable of the dragon which guarded the Golden Apples in the garden of the Hesperides has many parallels in our own country—one of which is that our apple is the fruit of the " forbidden tree " mentioned in the Scriptures; and the shaking or trembling of Aspen leaves is attributed by the Highlanders to the fact that it was from the wood of this tree that the cross of the Saviour was made.

There probably exists a good deal of popular misunderstanding concerning the various excrescences found on trees. "Oak-galls" are not plant productions, but are caused by minute gall-flies, and are designed as a protection for their tender offspring before they are able to shift for themselves. Oak-apples are excrescences of a like nature, divided into a number of cells, each of which contains a grub, a pupa, or a perfect insect, according to season.

The oak spangles which stud the under sides of leaves were formerly thought to be parasitic fungi, but are now known to be caused by Gall-flies, as are the tiny spheres attached to the flower-stalk. Myths of a like kind relate to the reddish-coloured protuberances of the leaves of the Lombardy poplar; while in reality, the tissue thus moulds itself at the working of another species of aphis. But similar popular delusions are almost endless.

CHAPTER V.

ANOTHER TALE OF ARCADY—I.

MY Northern Arcady of 1790 was unfortunately placed. It was away from the well-trodden tracks of commerce, and a hundred miles from the nearest manufacturing centre. The folk of the fell dales of a century ago lived out their lives in a narrow way—lives that were bounded by the hills of the valleys that had bred them.

One of the yeomen, more cosmopolitan in his ideas than his neighbours, and who had climbed to the highest peak of the hill range, returned and told his neighbours that England was a bigger place than he had thought. Yet these primitive fell folk were a fine race. They had the virtues which attach to dwellers in a hill country. On the whole, they were thrifty, industrious, and independent. Most of them belonged to the " 'Statesman " class, and both owned and farmed their holdings. Many of these had been held by one family for generations, for the 'Statesman was essentially conservative, and the world went slowly then.

The old-fashioned yeoman followed the most primitive methods of agriculture. He lived along the sides of the dales, a mountain stream rushing through his rich meadows immediately below. His house and barns, built primarily with a view to shelter, were composed of rocks and boulders from the fell slopes, and were more like productions of nature than of art. The homesteads were generally planted at the base of the mountains, as there the soil is richest and deepest. The valley bottoms make productive meadows; and although the fell sheep often graze them far into summer, they yield abundant crops of hay in July. In these remote dales, however, the summer months are often wet ones, and the hay harvest is much delayed. Taking our stand by the margin of the valley stream, we have, first, the meadow slip, then the " intacks " or fell-side pasture, the " grassing heads," and, finally, the mountains. Many of

E

the enclosed hill pastures are fringed with shaggy underwood and bosky dells, vestiges of primitive forests. In limestone caverns and recesses of the hills, remains of bears, wild boars, and wolves are found; while on one of the fells there still roams a herd of wild Red Deer.

A century ago, a large proportion of the land in Cumbria was owned by these 'Statesmen, the farms ranging in value from £40 to £50 a year. Many of them were held by one family for generations ; but too often, as they descended from father to son, they became heavily burdened with charges to the younger members. Mortgages and interest accumulated until the case of the 'Statesman became hopeless, and he was glad to find a purchaser for his little demesne. A series of bad seasons, loss of stock, or a prolonged winter, would not unfrequently prove the last straw ; or it sometimes happened that the yeoman's family became too large to be sustained by the estate. The natural result was that the small holdings were gradually merged in the larger ones, until now the process of assimilation may be said to be complete. The few that have survived have done so by consolidating small holdings where family interests were identical. In this way, there are numerous small farms worth £100, £200, £300 annually; while a few more extensive ones command a rental of from £500 to £700. Some of the large sheep-farms now embrace whole villages, and are from one to five thousand acres in extent. A number of small farms at from £60 to £70 remain, but these are comparatively rare.

In most cases the 'Statesman's holding was essentially a sheep-farm. He had right of "heaf" for four or five hundred sheep upon the Common. The times of lambing in spring, of washing in the fell "becks," and of shearing in late June, were among the events of the year. The sheep were of the hardy Herd-wick breed, climbing to the bleak fell-tops at the coming of the snow; and in this was their safety. They were rarely buried in drifts, and were clever at scratching away snow to get at the hidden herbage. The 'Statesman's daughters spun the wool of his own sheep, and from this the clothing of the family was made at home. Spinning-wheels are occasionally still used, but are oftener found stowed away in the lumber-loft.

At the time of which we speak, the internal
communication of the northern counties was
bad indeed. The roads were so narrow that
only pack-horses could travel along them, and
in this way the traffic of the dales was conducted. Carriage roads there were none, and
it was probably owing to this fact that, just as
the yeomen spun their own wool, so they grew
sufficient grain to last them through the year.
Marks of the plough are often to be seen on the
Commons and moorlands; and in these marks
we may read one reason of the rapid decline
of the 'Statesman class. About the beginning
of the present century thousands of acres of
the lower-lying Commons were enclosed. The
Continental wars then raging had sent up all
kinds of grain to a price unknown before. The
yeomen reaped a rich harvest; fresh land was
broken up, and some of it yielded enormously.
Every available bit of land was ploughed, and
corn crop after corn crop taken off. These were
flush times for the 'Statesman, and lavish habits
were contracted. Peace came, and brought its
natural consequences.

The spell of agricultural depression from
which we are still suffering has dealt the death-

blow to the 'Statesman. Ten years ago one mountain dale at least was wholly tenanted by farmers of this class. To-day not one remains. In this quiet spot they had long lived uneventful lives, fairly prosperous, and content. Their little world was small, but their wants were few.

As many of their sons and daughters as could be spared went out to hired service, or if one of the lads happened to be a weakling, he " was bred a scholar." At the beginning of the present century scores of yeomen's sons held small " livings " in their native northern dales.

The life of the fell folk must have been terribly lonely in winter. They rose and went to bed with the sun, their only artificial lights being made from rushes and mutton-fat. Among the shippons in winter these were carried in old-fashioned horn lanterns, all of which articles they manufactured by their own simple methods. There were no markets for milk and butter, and so the former was converted into cheese, mostly of a very low quality. Oatmeal, salt-beef, bacon, and dried mutton, constituted the staple food of the better class yeoman, whilst in summer eggs, veal, and other articles of diet were added. In the neigh-bourhood of the provincial towns salmon was sold at twopence a pound, and other fish food was equally cheap. With almost every article of diet, however, oatmeal was largely eaten. But this state of things has long since come to an end.

The great revolution in the manners and customs of the 'Statesmen began at the close of the last century. Much that was singular and characteristic among them vanished at the opening of the turnpike-roads. This and the extension of commerce afforded facilities for the purchase of objects of elegance and ease, and produced results which soon spread even to the more secluded mountain dales. The tracks of the pack-horses were difficult at all times, and, as a rule, were ill-kept. Personal inter-course with the southern portions of the country was extremely difficult. We speak lightly now of the will-making of our forefathers before starting on a long journey. To them it was a matter of no light consideration, and those whose business took them from home settled their worldly affairs before starting. Many of them were manufacturers of woollen goods, and these they themselves sometimes travelled to sell.

With macadamized roads the strings of pack-horses began to decline, and post-chaises were introduced. This was in 1754, and three years later carrier's carts and waggons had come in. In 1763, the first stage-coach was seen in the North, and was drawn by six horses. It ran from Edinburgh to London, and took four-teen days for the journey. This was express speed, so to speak, and the Edinburgh professors of that day warned persons from travelling by the wild and whirling vehicle, as the rate at which it went would bring upon them all manner of strange disorders, chief among which was apoplexy.

This, as already stated, was the beginning of many and rapid changes. Before that time spinning-wheels were in every country-house. In most cases the wool was taken from the backs of the manufacturer's own sheep, and each process, from first to last, was performed by some member of the household. Every woman was a first-rate knitter, and there was a vast "output" of old-fashioned, blue-grey stockings. Outside the domestic circle a few

hand-loom Weavers wove cloth of duffel grey
for the men, and russet for the women. Finer
wool for finer work was carefully combed
within the settle-nook; but almost all for domestic use. The Wool-comber was a great man
in those days, and the itinerant who tramped the
country and turned wooden dishes and such like
articles was always welcome. It was from observation of such men as these that Wordsworth
drew the character of his immortal " Wanderer."
The produce of a field of flax now yielded
material for holiday attire, and was in great
request among women. The travelling tailor
went from house to house in search of employment, and in the larger of them was sometimes detained for weeks. He worked for daily
wages, and amply paid for his meat and drink
by the news he brought. In those days news
was news—it could not be had for a half-
penny.

Although home produce was large, money
was scarce, and the earnings of servants were
paid "in kind." The girls received shifts and
gowns and aprons; the men, shirts and coats,
with sometimes a little wool and corn.

In money, the annual wages of a man-servant

was about £5; of a woman £2 5s. In our old hill church is a fair mural monument to Dame Gylpine, of The Hall. I have before me an account of "The holle yeare waigs of alle her servants at Sellsatt," and these amount to 290s. for eight men and nine "maydes." In the days of Dame Gylpine, a thorough knowledge of domestic art was considered necessary to fit a lady for the duties of wife and mother.

Where one girl can knit a pair of stockings now, there were a hundred then, though in the matter of music and philosophy the proportion is reversed.

In bygone days the gentry of the countryside deemed it no degradation to manage their affairs, even down to the minutest detail. Iron-shod shoes (clogs) were generally worn; the parson's and yeoman's children appearing in them at church on Sundays. But these clogs proved injurious to the wearer's stockings, and it is said that careful housewives used to smear the heels of the latter with melted pitch, and dip them immediately in the ashes of the turf fire. Fixed in the woollen texture, the mixture became both hard and flexible, and was well adapted to resist friction.

Flax has long since ceased to be grown in the field; there is now no hempen cloth, and the old methods of spinning have gone out. Into some of the northern valleys, Mr. Ruskin—striving after an unattainable idea of pastoral peace and happiness—has re-introduced spinning-wheels; but the whirr of them speaks most of his eccentricities. Among the manufactures of the North were yarn, hose, horn lanterns, and coarse druggets; but they have long ago been supplanted. Even the poorest have turned their backs on honest homespun, and now trick themselves out in webs of draggled embroidery. The old " stuffs" are gone, and materials with greater gloss and less substance have taken their place.

Up to the beginning of the present century, which constituted an era in the history of the dales, the domestic economy of the 'Statesman class was in a backward condition. Their houses were ill-contrived, and hardly in keeping with modern notions of decency. The water supply was, of course, indispensable; but, instead of digging a well or conducting water to their homesteads, these were invariably built by the sides of the fell " becks." Consequently, many of them stood in moist situations; and occasionally, in autumn, foaming torrents tore up the folds and washed away the outbuildings. These, like the houses, were low, and a man of ordinary height could not pass the door-lintel without stooping. The floors of the houses were below the level of the ground without, and entrance was made by a descent of one or two steps. The basement was divided into three apartments—the " buttery," which constituted the general larder; the common hall or kitchen, which formed the living room; and a slightly raised chamber, in which the master and mistress slept. The whole was either rudely paved with cobbles from the river bed, or had a floor of flattened loam.

There was no fire-grate, nor is there yet in many of the smaller dwellings of the dales; the peat or wood fire being laid on the hearth. The fires were " raked " at night, and some are known never to have been extinguished for a century. " Raking " was easier than having to re-kindle fuel with the aid of flint and steel, and was the universal practice. The chimney-place was one into which a waggon might have been driven, being twelve feet or

more in diameter, and open in front. In the
funnel there hung joints of beef, mutton, and
pork, while sometimes a dozen hams were
smoking in the chimney at a time. A long
sooty chain ending in a crook went up from
the fire to a beam above, and this bore the
heavy iron pans of the period. Many of these
were supplied with a funnel, by which to let
the steam escape ; and, as there was nothing in
the shape of ovens, all food was cooked by their
aid. At all seasons of the year sooty water
trickled down the wide chimneys ; and the
members of the household sat and went about
their domestic duties with their heads covered.

The second story of the house, called the loft,
was open to the rafters, and constituted the
sleeping apartment of the servants and children.
In most cases there was but one chamber,
undivided by partitions, and here the depen-
dents were lodged—the men at one end, the
women at the other. Beyond a rope stretched
across—upon which coats and gowns, articles
of both male and female attire, were hung pro-
miscuously—there was no division whatever.
In the houses of the tradesmen of the adjoining
provincial towns, where the custom was to

provide lodgings for journeymen as well as apprentices, matters were even worse.

The furniture of these northern homes was rude both in design and execution, but it was useful and homely, and eminently in keeping with the houses that contained it. The one quality that was striven after in all domestic utensils was serviceableness. Almost everything was of wood, pegs of this substance invariably supplying the place of nails. Wooden latches were to all the doors and windows, iron being almost unknown in domestic architecture.

One great feature of the farmhouses was their arks and chests. These were curiously and quaintly carved, with carvings " all made out of the carver's brain." Coleridge lived in a district where the work of the home-bred carver was everywhere to be seen ; and doubtless the line in " Christabel " was so suggested. In the arks were kept oaten cake, malt, meal, preserves, and dried meats. The " chest "— hoary, heavy, and tall—contained the clothes of the family, often with an immense quantity of linen and cloth of home manufacture— mostly the work of the women. These stores were largely drawn upon for bridal dowries.

Old china, dress material in flowered silk or satin, and a few pieces of plate—family heirlooms—found a place in the chest.

The huge bedsteads in use were of massive oak, with testers of the same substance. The chairs were generally made of clumsy wainscot, and some were fashioned from the trunks of hollow trees—the carpenter completing what time had begun. For table, there stood in the common hall a board of from four to seven yards in length ; the rude stand being furnished with forms or benches along its sides. Upon these the family and guests seated themselves at meals. Maple trenchers supplied the place of plates, and liquids of every description—milk, broth, beer—were served in wooden vessels, made with staves and hoops.

The protection from cold and wind—which freely found their way into the common hall by the chimney, as well as the badly jointed doors and windows—was provided against by a screen placed in front of the turf fire. In the centre of the hearth stood a square, upright staff, having a row of holes along one of its sides, its lower end fixed into a stout log of wood. This simple contrivance supported the candlestick, which was thrust at convenient height into one of the holes.

In the warm, though smoky retreat of the settle-nook, the family spent the long winter evenings in knitting, spinning flax, carding wool, and other home industries. Conversation at such times, especially that of the elders, had but one result—perpetuating the credulity of the times. The talk constantly ran on apparitions, omens, workers in witchcraft, and more innocent fairy tale.

But at the beginning of the new era these things obtained less countenance, and a general change for the better began. Provincial newspapers were started and wonderfully enlarged the narrow world of their country readers. Innovation came steadily from the South, and the rude artisans were ousted. Old handicrafts were subdivided; the cabinet-maker invaded the province of the carpenter; the worker in metals that of the maker of wooden platters; and the great army of itinerants rapidly declined. The transformation during the period indicated exceeded that of any century that had preceded it in the North.

CHAPTER VI.

ANOTHER TALE OF ARCADY—II.

As yet we have but casually mentioned the actual farm life of the northern yeoman. At the time which this chapter concerns, their land operations consisted in the very oldest methods. The people rarely migrated from one valley to another; they had few wants; and of them it might be said that there the richest were poor, and the poor lived in abundance. It is a remarkable testimony to the practical shrewdness of northern farmers, that whilst a century ago their farming was of the worst possible description, the depression in agriculture to-day is less felt among them than perhaps in any other portion of the country.

Theirs was the "old system" of husbandry. When grass land was broken up, it was sown with black oats, all the available manure of the little " estate " being bestowed upon it for the succeeding barley crop. The third year the land was laid down again to fallow with a second crop of oats, but always without grass seeds, so that the future herbage came no one knew exactly how. In such case, however, nature seemed to let loose her ubiquitous weeds, and soon a green mantle overspread the fallow.

One of the first improvements upon this state of things was the application of lime to such lands as were wet and moss-grown. This process was universally ridiculed until the result was seen, when lime-kilns sprung up everywhere.

Although spring-wheat was cultivated in the northern counties as early as the sixteenth century, green crops as food for cattle are of recent date. As to pot-herbs and the produce of the vegetable garden generally, a century ago they were nearly unknown. Oaten bread, dressed barley, and onions constituted the more cooling diet of the common people, with very little variety. At the middle of the eighteenth century, however, Common-gardens were laid out in the environs of most northern country

F

towns, and at the same time the culture of fruit-trees became general. This was an important step, for not only did it supply a needed article of diet, but it was the beginning of a new industry. The more hardy fruit-trees were peculiarly suited to the humid valleys of the North, and in time yielded enormous crops.

As yet the art of fattening cattle was but little understood; the first experiments in this direction being tried upon sheep in winter. This was an important matter, and raised great hopes in the minds of breeders; for as yet the winter supply of animal food had proved wholly in-adequate. The stock fed in autumn was killed off by Christmas, and, with the exception of veal, scarcely any fresh meat appeared in the markets before the ensuing Midsummer. This dearth was provided against by the more substantial yeomen and manufacturers by curing a quantity of beef at Martinmas—part of which was pickled in brine, the rest dried in the smoke of the capacious chimneys. On Sunday, the farmers' wives boiled a huge piece of meat from the brine-tub, which on that day was served hot. From that time as long as the joint lasted it came up cold, relish being given to it by the addition of oatmeal puddings. Hogs were slaughtered in great numbers between Christmas and Candlemas; the flesh being converted into bacon, which, with dried beef and mutton, afforded a change in spring. The only fresh provisions in winter consisted of eggs, poultry, geese, and ill-fed veal, the calves being carried to market when only a few weeks old.

What is here set down has reference to the small farmers and better-class yeomen. The class next below them knew but little of their comfort, and nothing of their luxuries. The artisans and the land-labouring classes were badly housed, and were fed on barley boiled in milk; with the addition of meal-bread, butter, and a small quantity of salted meat. This diet told terribly upon the poorer population in spring, for ague set in with painful regularity. The culture of esculent vegetables became more common, and potatoes began to be generally, though sparingly, used in 1730. The cultivation of this root operated healthily upon the inhabitants, and made them much better off than when they were wholly dependent upon grain. About this time parcels of tea began to be received in the North from the Metropolis ; but, in ignorance of its use, it was smoked instead of tobacco, made into herb puddings with barley, and, as might well be supposed, was rather long in becoming quite popular.   As the use and virtues of this foreign luxury became better understood, neatly turned cups and saucers of wood made their appearance, these being commonly used instead of porcelain.

The consumption of wheat became greater and more general after tea was introduced, though at the beginning of the century flour made from this grain was never seen in the cottage of the labourer, and rarely made its appearance on the tables of the middle-class except on days of festivity.   Frugal housewives evinced their attachment to economy on these occasions by making their pastry of barley-meal, which they veneered with a thin cake of flour.   A curious custom survived until recently from out these bygone things that shows the estimation in which wheaten bread was held by our ancestors. A small loaf was thought a fitting gift from the dead to the living, and every person who attended a funeral received one at the door of the deceased, and was expected to carry it home. The increased demand for wheat encouraged the hill farmers to pursue the more modern methods of cultivating it ; and though the experiment has everywhere succeeded, thin cakes of oatmeal without leaven are even now daily eaten in quantities in almost all the houses of the mountain dales. It is upon this and oat-meal porridge that the famous wrestlers of Cumbria are reared, and at present there seems but little likelihood of its disuse.

And now, to come to another side of the subject—to more recent times. It is hardly remarkable to those who know the 'Statesman class, that its descendants of to-day have suffered less from recent agricultural depression than others. And this, perhaps, is owing to the sterner virtues and natural shrewdness which are among the characteristics of the dalesfolk. At the same time, it must be understood that nearly all the farmers' sons in the north go out to work as farm servants ; and it may at once be said that the northern farm labourer is nothing akin to his southern brother. It is probably no exaggeration to say that he is superior to him in every way in which comparison is possible. The southerner seems unable to lift himself

above his surroundings, whilst the northerner
almost invariably strives to do this, and not
unfrequently succeeds.

As a lad he is well schooled, and at four-
teen begins "service." At half-yearly hiring
—Whitsuntide or Martinmas—he goes to the
nearest country town and stands in the market-
place. He is attired in a brand-new suit, with
a capacious necktie of green and red. These
articles he has donned upon the memorable
morning, and, as a gift from his parents, they
constitute "his start in life." The country
barber has left his head pretty much as the
modern reaper leaves the stubble, and has not
stinted him of grease for his money. As an
outward and visible sign of his intention, he
sticks a straw in his mouth and awaits the
issue. For the first hour or so he keeps his
eyes bent on the pavement, as though to read
the riddle of his life there, but presently gains
confidence to look about him. After waiting
the greater part of the morning, and seeing
many of his fellow-men and maid-servants
hired, he is accosted by a stalwart yeoman,
who inquires if he wants a "spot" (a place,
a situation). The boy replies that he does, that

he is willing to do anything, and will engage
for £5 the half-year. A bargain is soon struck,
and the stalwart urchin from the "fell-heads"
wanders off to lose himself in the giddy gaiety
of the Fair.

If, ultimately he likes his "place," and is
well and kindly treated, we may not see him
again at hiring for a couple of years. During
this time he has made himself generally useful,
has become a good milker, and has shone conspicuously at hay and harvest. He has proved
himself a "fine lad," too, and has had his
wages raised by way of reward. At twenty
he is stalwart enough to hire as a man; and
now his wages are doubled. He asks and
obtains £18 for the year, or even £20, if he be
entering upon the summer half.

The farm servants of Cumbria "live in,"
and have all found. They are well fed, well
housed, and take their meals at the master's
table. But, if they are well fed, they are
hard worked, and in summer often rise as
early as three or four in the morning. In
these counties, which constitute a vast grazing
district, the labour of the farm servant is much
more general and interesting than that of

the southerner, who works on arable land. Physically, the northern man is much superior, and is generally an athlete. He is come of that stock which so stubbornly fought the Border Wars, and now excels in wrestling, mountain racing, and following the wiry foxhounds among the hills. Of course, these sports only offer during the scant holidays in which he indulges, or in mid-winter, when there is little to be done beyond tending the stock. It is necessary that the northerner should be a big, sturdy fellow, since the ground which he has to work is rough, and he has not many helps in the way of machines.

A few years ago, and during the flourishing times of agriculture, the northern labourer obtained from £35 to £40 per annum, still, of course, "living in;" a few picked men could even command £45. Now, however, that we are come upon times of depression, the best men are glad to work for the first-named sum. In proportion, girls are much better off in the matter of wages. There is probably less competition among them, owing to the fact that there is a great temptation for country girls to migrate and enter service in provincial towns.

Here they are not so hard worked as in the farmhouses, and have the satisfaction of being engaged in what they esteem a more "genteel" occupation.

Many of the men, when about thirty years of age, are able to take small farms of their own. Nearly all the 'Statesman's sons do this, probably without any outside help. I know a man who saved £120, which sum he divided and deposited in *three* banks. This was his whole wealth, and he explained to me that he did not want to lose his hard-earned savings if the banks should "break." His object was to acquire a small farm. He has now succeeded.

From the fact of "living in," as nearly all the farm servants of Cumbria do, it need hardly be said that early marriages are rare. All the better men look forward to the time when they can have a farm of their own; and as soon as they have obtained a "holding," they look out for a wife. This fact alone speaks well of their thrift; but it has its dark side. How far the two things are connected may be a matter of speculation, but it is notorious that the number of illegitimate children in

Cumbria is far above the average, and most of these are undoubtedly born of the agricultural classes. The Registers of the country churches prove this. Still, it is pleasing to be able to record the fact that in the Dales, sooner or later, those who have been wronged "are made honest women of" by marriage.

In the remote dales the farm servants are as conservative as their masters. Not only in politics, but in their whole surroundings. At one time of the year the sledding of the peat constitutes a considerable portion of their work. Nothing but sticks and peat are used for fuel; and the peat has to be "graved," then stacked, and finally brought from the moorlands on sledges. This is done in autumn. Spring is occupied in tending the mountain sheep, the time of lambing being a particularly busy one. The sheep-washing is also pleasant, and shearing, a month later, brings quite an annual festival. The "clipping," where the holdings are essentially sheep-farms, is one of the great events of the year. Before the days of dipping, salving sheep came in late autumn, and brought a time of terribly hard work. The process was slow, and sometimes a thousand head had to be gone through. To enable the men to do this, they had candles fixed in their caps, and worked early and late. Salving is now superseded—a thing of the past.

In Cumbria, winter, except unusually severe, is hardly a busy time, for then most of the sheep are brought from the fells and wintered about the farm. It occasionally happens, however, that these are buried deep and have to be dug out, or that trusses of hay have to be carried to the fells. In fact, half the time of the northern farm labourer is taken up in connection with the sheep.

Many of the men of whom I speak are going south as Hinds; and men from Cumbria are as a rule preferred for such situations, as the advertising columns of the agricultural journals show. They are practical farmers, shrewd, and not afraid of right-down hard work. Whenever they go south, they endeavour to graze more, and plough less land; and so follow in the lines in which they were reared. And it is generally admitted that they are successful. Just now many of them are obtaining southern farms which have gone out of cultivation, and some have already

shown what can be done in this way. They
pay but little rent, and will doubtless do well
with their bargains. In short, the Cumbrian
farm labourer, wherever he appears, literally
drives his weaker brother out of the field, and
he has already left his mark upon southern
grass lands.

This is the man who is the outcome of
the old 'Statesman, who is now nearly extinct.
On his native heath he is man instead of
master, and may be happier for the change.
His ancestors have been revolutionized out of
existence. His customs, his clothes, his food,
his mode of living are changed. As a peasant
proprietor he has gone to the wall.. Many of
his mountain folk are now included in the pale
of humanity, which certainly could not have
been said of them a couple of centuries ago.

For many of the facts in this chapter the author is in-
debted to a tract by a local Antiquarian published in 1847.

CHAPTER VII.

Of British field-sports, perhaps "wild-shooting" is the most fascinating. This can only be pursued over wild though fairly well-stocked ground, with the prospect of a miscellaneous though always uncertain bag. It is this latter element that gives such zest to sport. On the last day of December a light fall of snow had thinly carpeted the open woods. Never was morning more beautiful! The feathered rain was crisp to the tread, and the noon sun converted the atmosphere to that of summer. The water in the bay was blue; the snow-peaks of the hills rose-tinted; and snow-crystals shot up over the landscape and made it gloriously, dazzling bright!

We were three — old Phil, a broken-bred

spaniel, and myself. The day and its anticipa-
tions were just such as we could drink in to
the fullest. We knew and mutually understood
each other, so that each would work to the
gun of each, and the dog to both.

Our way lay over the snow, through low
scrub of bright chestnut oaks, glowing with
colour. Picking the acorns from beneath the
trees, a brace of Pheasants went away. Both
birds were bagged, though after the towering
cock I had to send my second barrel, as a
badly broken wing did not at first stop him.
The hen fell dead to Phil as it topped the
bushes. There is no gainsaying the beauty of
the pheasant of our woodlands, and, with the
frost, the iridescence of the plumage is perfect.
A flock of burnished Wood-pigeons flew wildly
from the beeches, though far out of range.
Then we have a spell of sharp shooting.
Rabbits rush across the rides, and are rolled
over as they pass through. Saplings of oak
and hazel do not stay our firing, the shot
cutting through these and killing beyond. For
a time, on the hillside, sharp shooting is com-
pulsory, and a dozen more rabbits are added to
the bag. Then we emerge from the coppice and

pass under the pines, walking over layer upon layer of pine-needles. A Long-Eared Owl floats from out the gloom; but the winged mouser is allowed peacefully to skirt the plantation.

Almost all life and vegetation are blotted out here, and the aisles are silent. The hoarse croak of a Carrion Crow comes from the rocks by the bay, and then, disturbed, passes high overhead. The pine-wood passed, we emerge to light and sunshine. A Woodcock, with its peculiar owl-like flight, rises from a bracken bed, and contributes its long bill to the bag. A Sparrow-hawk sails silently over the trees, but its sharp eye detects us, it wheels, and is soon out of sight. Another flock of wood-pigeons is more successfully stalked than the last, and a pair added. Along the edge of the wood, and among the still remaining stubble, some pheasants are feeding; and as these get up another brace is added.

Then comes an hour's stalk. A covey of Partridges is put away time after time, but always far out of range. Marking the birds, we follow on; and just as we persist so the birds become wilder. Soon, however, the ground favours us; the covey drops over a

brac a hundred yards away, and, following,
we come right upon them. Selecting our birds
on the outskirts of the flight, we have our
reward by bagging three, each of us emptying
both barrels.

Having dealt out this swift retribution, we turn
to the Snipe. We find these, as we expected,
in a marshy meadow at the bottom of the wood.
Five are killed with seven shots, many others
getting up out of reach, and calling " Scape!
scape!" as they fly off. It was remarked that
the brace that got clear away were fired at
just as they rose. The surest method in snipe-
shooting is to exercise patience, and pull just
when the bird has finished zigzagging. The
finger must rest lightly on the trigger, however, and not a second be lost, as by this
time the bird is just passing out of range. It
is marvellous how much thought can be exercised in a moment whilst shooting! Even as
the eye glances along the barrels, and the finger
presses close with every instinct to pull, some
slight circumstance intervenes, entailing a train
of thought and its deductions, which ends in
letting the game go.

From the rushy marsh a Woodcock is bagged,

and here, through the turnips, a Hare comes lopping by the fence. We stand motionless as it makes towards us in a series of lateral leaps. It nibbles the herbage in its track, and then, getting wind of us, stops, stamps, and rears itself on its haunches. In the very act of bounding away it rolls over stone dead.

A sentinel-like Heron, which stands gaunt and motionless on one leg in the icy stream, rises and flaps over, but is allowed to pass. This lonely fisher of the tarns and streams has now enough of cruel want without being shot at. Over the next fence is a secluded sheet of water, which usually harbours ducks. This we approach stealthily. A shot from each brings down a brace of Teal as they rise from the reeds, though we restrain our hands at a Kingfisher as she shoots up the beck. A pair of Dippers paddle along the pebbles, and, following the progress of the stream that feeds the tarn, a Golden-eye Duck and a great Black-backed Gull reward us. The runner brings us down to the green rocks by the bay, under which we intend to lie and wait for the in-coming birds. The white foam of the tide creeps on in the distance, and with it the wild

G

sea-mews. All the birds that line the channel
now grow active, and either run or flap along
the lagoons. Five ducks fly low, coming from
seaward, but keep so far away that we cannot determine the species. As the tide rolls
to our feet, thousands of sea-birds fly crying and
calling over the waters. A Woodcock looms
out over the sandy tract below, and is shot. It
is followed by another, and another, until five
are killed. This is wholly unexpected, both
time and place being unusual; and then five
Woodcocks in company are rare.

Presently the tide ebbs; the cockle-beds
teem with countless birds; and with the twilight
the ducks begin to arrive over their feeding
ground. We are now snugly ensconced, with
the wind behind us; a cavernous recess in the
rocks affords a seat; and from the promontory
we command several likely pieces of water.
Nor have we long to wait, for the spaniel
quickly rises to its feet and stiffens every
muscle. We catch now what his quick ear
detected before us, and a quacking comes
from overhead. Four Teal wheel around,
then make up channel. A single black-look-
ing duck comes within forty yards, and falls

to my gun; its mate, which we had not previously seen, rises, and is brought down with
my second barrel. Phil next kills a Mallard flying high over, and then for a while
nothing is doing. The birds are left where
they fall, and retrieved after the shooting.
The black ducks eventually prove to be Scaups.
The fishermen hereabout call them "dowkers"
and "bluebills," and take great quantities in
their nets.

Even in the dusk we can make out the white
plumage and chestnut saddles of a flock of
Shelldrakes as they "plump" right down
into the water. They soon settle to feed, and
offer a splendid chance. They are evidently
unaware of our presence, and, firing four
barrels, only one escapes. Two Teal and a
grey duck are added, and although the birds
still continue to come in from the sea, our
bags are full; cold and benumbed, we creep
from out the rocks and plod home over the
heavy sand. Ducks and shore-haunting creatures get up everywhere before us, and a deer
has come down from the park to lick the saline
stones. The night is starlit, and strings of
ducks pass over the sky-line. Once we hear

the gaggle of wild geese, but the sounds appear far up.

Such is wild-shooting where game is fairly plentiful. It is thoroughly hard work, though in the exhilaration and excitement this is never felt. Our bag is a heavy one—at least to the shoulders—and its contents are as miscellaneous as the most ardent sportsman could desire.

CHAPTER VIII.

THE Gamekeeper's Cottage stands at the end of
the Oak Lane. An orchard surrounds his
dwelling, the brown boughs now drooping with
ripened fruit. Under an overhanging sycamore
is a kennel of silky coated setters and a brace of
spaniels. The former have beautifully domed
heads and large, soft eyes. The spaniels with
their pendulous ears are a black and a brown.
Pheasant pens are scattered about the orchard,
each containing half a dozen birds. In a disused shed are traps for capturing game, and nets
and snares found in rabbit runs or taken from
poachers. The keeper does not always take these
engines when he finds them, but waits quietly
until they are visited by the " moucher;" and
then he makes a double capture. Few of the

poachers, however, leave their traps after dark,
and only the casual is caught in this way. At
the other end of the orchard, divisional boxes
are ranged round an old barn-like building
where pheasants' eggs are hatched. A shaggy
terrier, with fresh mould upon its nose, peeps
from beneath the shed doorway. Drowsy blue-
bottles buzz about the vermin larder, and under
the apple trees are straw-thatched hives. Contented pigeons coo and bask on the hot slates of
the barn-roof, and bird sounds are everywhere.
These blocks, upon which sit their falcons, act
as a reminder of an old English sport fast
passing away. They are Merlins and Peregrines, kept by the keeper for a friend, who is
fond of hawking. The merlins can pull down
partridges, while the peregrines are flown at
larger game. No sport so exhilarating as
falconry; none so fascinating!

The interior of the Keeper's Cottage is as
characteristic as its surroundings. Here are
guns of every description—from the old-
fashioned fowling-piece and matchlock to the
ponderous duck-gun. Above the chimney-
place hangs a modern breech-loader with
Damascus barrels. The Keeper admires the

delicate mechanism of this, but depreciates the spirit of the age which produced it. Such cunningly devised engines will make old-fashioned sport, or what he calls "wild-shooting," extinct.

Against the walls are cases of stuffed birds, with a red Squirrel or a white Stoat to relieve the feathers. In one case a knot-hole is imitated, from which peer three young Weasels; and an old one is descending the bole with a dead bird in its mouth. All these are portrayed to the life by the keeper's own hand. Looking at the contents of the cases, he deplores his want of ornithological knowledge in earlier years. Among the stuffed specimens are a Greenland Falcon, a pair of Hobbies, several rare Owls, a Kite or Glead, a Rose-coloured Pastor, and others equally rare.

The Gamekeeper's life is essentially an outdoor one. He is far from populous towns, and needs but little assistance. Poachers rarely or never come to his preserves in gangs, and a couple of village "mouchers" he can manage easily. His powerful frame has once been the seat of great strength, though now it needs but a glance to show that his eye is less keen, and his hand less firm. Still he is quick to detect, and with his hard-hitting muzzle-loader he rarely misses. Given favourable conditions, he is almost infallible with the gun, though he gives his game law. He cannot now cover his extended ground in a single day, and perhaps does less night-watching than formerly.

His beat covers a widely diversified district, with almost every species of game. The Pheasants wander about the woods and copses; the Partridges are among the corn stubble; and rabbits pop in and out everywhere. Hares haunt the meadows and upland fields, and Snipe go away from the marshes. Woodcock come to the wet woods, and a host of sea-haunting creatures feed along the shore. There is a Heronry in the wood, and pigeons build in the larches. Of the habits of these the Keeper is full; and if he is garrulous he is always instructive. By daily observation he has found that animals and birds have stated times and well-defined routes. Exactly at the same hour, according to the sun, the Partridges and Pheasants resort to the same spots. Hares follow their tracks day by day, and Rooks fly, morning and evening, along the same valleys. Nightly,

Herons stalk the pools, and the Otter traces the mountain burns to their source. At noon a Sparrow-hawk speeds by the covert, and at evening a Kestrel hangs over the rick-yard. In the afternoon, regularly, Weasels run along the old wall; and, as these things, the flowers in their times of opening and closing are not less constant.

The Keeper's domain encloses a park in which are Red Deer and Fallow. Sometimes he has to shoot a fawn for the "great house." This he singles out, hitting it if possible just behind the shoulder. In season he must provide a certain "head" of game. Twice weekly he procures this, and takes it to the Hall. For its proper hanging in the larder he is responsible. When he wants game, he knows to a yard where it may be found—where the birds will get up, in what direction they will go away. If a Hare, he knows the gate or smoot through which she will pass, and out of this latter fact he makes capital. It is well known to poachers and others that when once a hare has been netted there is no chance of its being retaken in like manner. Rather than go through this a second time, even though a lurcher be but

a yard behind, it will either " buck " the gate
or take the fence. Consequently the Keeper
has netted every hare on his ground. This
greatly reduces the poachers' chances, and wire
snares are now the only engines that can be
successfully used.

Spring and summer are taken up with
pheasant-breeding, and this is an anxious
time. The work is not difficult, but arduous.
And then, so much of the Keeper's work is
estimated by the head of game he can turn
out. This result is palpable,—one that can
be seen both by master and visitors. There
is nothing to show for long and often fruitless
night-watching but rheumatism, and so the
Keeper appreciates all the more readily the
praise accorded him for the number of well-
grown birds he can show at the covert-side.

After pheasant-shooting in October, the
serious winter work of the keeper begins. Each
week he has to kill from three to four hundred
rabbits, which are sent to the markets of large
manufacturing towns. He can employ what
engines against them he pleases, but the number must be produced. The work being long
continued, becomes monotonous. Firing a hun-

dred shots a day is now more jarring than it once was, it has made him slightly deaf, and he adopts other means of destruction. He works the warrens in winter, but long waiting for a glutted ferret in frost and snow is not pleasant. Under favourable conditions, however, a great many rabbits may be taken in this way. Iron spring traps are used in the rabbit tracks, but these are impracticable on a large scale; and pheasants and partridges, which run much, are apt to be caught in them. Moreover, it is now illegal to set these traps in the open.

The most certain and wholesale method of capture is by the "well-trap." This is a pit, placed immediately opposite to a hole in a fence through which the rabbits run from woods to field or pasture. Through this "run" a wooden trough is inserted; and as the rabbits pass through, the floor opens beneath their weight and they drop into the "well." Immediately the pressure is removed the floor springs back to its original position; thus a score or more rabbits may often be taken in a single night. In the construction of these traps rough and unbarked wood is used, and even then the rabbits will not take them for weeks. Then they become familiar, the weather washes away all scent, and the "well-trap" is a wholesale engine of capture. The rabbits of course are taken alive. These the keeper stretches across his knee, dislocating the spine. English rabbits are degenerating in size, and the introduction of some of the continental varieties would be beneficial. With the rabbits in autumn great quantities of wood-pigeons are sent away, the birds at this time becoming exceedingly plump and fat. An almost incredible number of acorns may be found in the crop of a single bird when the former have fallen.

These are a few of the Keeper's duties. He himself has a russet, weather-beaten face, bounded by silvery hair. He might stand for a picture of a highly idealized member of his class. So secluded is his cottage, that he locks the door but once a year, and that on Christmas Eve. He can remember when there was larger game than now—when Badgers and Wild-cats were not uncommon. One of his ancestors was an inveterate deer-stealer, as the parish books show. Then, the Red Deer roamed almost wild on the fells. To-day he has but one regret—

that he was not contemporary with the Wolf, the Wild Boar, and the Bear. Of these in Britain he has read an account, as well as of the vast primitive forests through which they roamed.

Just without the darkened shadow of the pine-wood is a sunny glade. The rides of the forest converge upon it, and here centres much of the life of the district. Delicious it is to lounge there on a long summer day, lying under the cool shadow of the shrubs. This have we often done with the Keeper, our employment being to watch the young pheasants among the scrub and brown uncurling brackens.

So long as the Gamekeeper can keep the young pheasants under his eye they are comparatively safe, but just now they are apt to wander; and when once they begin to do this there is no retaining them. Although fed daily with the daintiest food, the birds, singly or in pairs, may frequently be seen far from the home covers. Both man and nature's poachers know this, and are quick to use their knowledge. It by no means follows that the man who rears the pheasants will have the privilege of shooting them. At this season the birds take daily journeys in search of beech-mast, acorns, and blackberries, of which they consume great quantities. When the wandered birds find themselves in outlying copses in the evening they are apt to roost there. And herein lies the danger. The birds perch in the low nut-bushes and underwood, and are open to a whole host of enemies. The Sparrow-hawk, flying its beat at late afternoon, makes a swoop into the bushes and strikes down its prey; and a little later, as twilight comes, the brown Wood-Owls do the same. For months it has been the Keeper's chief concern to keep these birds under. And if they are destructive now, they were a dozen times more so then.

After hatching, and when the birds were transferred to the coops, the Keeper and his assistant spend the long summer days in feeding and guarding them from the falcons. The men lay hidden in scrub of oak, and birch, and hazel, and watched the young pheasants in the green rides. Small woodland birds swarmed everywhere, and fed among the pheasants; but at the warning cry of the Blackbird all the feathered throng dropped

G E. Lodge

L.Hutchinson lith   West Newman imp

The end of a Sparrow hawk

down into the shelter of the leaves, and a dark
shadow would glide over the sunny sward.
Then from out the pines there was a rush of
wings, and a Sparrow-hawk would be seen to
dash from the bushes with something in its
talons. This was repeated day after day, until
one afternoon as the hawk rounded the corner
of the wood, it was seen by the Keeper. He
lay close among the brush, but not so closely
as to escape the sharp eye of the hawk. It
doubled, but just as it did so the eye of the
old man glanced along the barrels, and his
finger touched the trigger. There was a
puff of white smoke, a cloud of feathers, and
the marauder dropped with a dead thud to
the sward. The old man picks up the bird,
follows a path through the bushes, and suspends
it in his " larder." This consists of a number
of unbarked rails nailed across two stalwart
oaks. Here is a sorry array of Cats, Jays,
Magpies, Squirrels, even the skins of Vipers
and Slow-worms.

The Sparrow-hawk is the most arrant of the
poachers. Ask the Gamekeeper to detail to
you the character of this daring marauder,
and he will record a black and bloody list of

depredations against the "spare"-hawk. He
knows but little, however, of the laws which
govern the economy of nature; and if he did,
or would, what are they compared to the head-
tax for those he can display on the vermin-
rails? The freshly added Sparrow-hawk is by
no means the only one of its kind, for there
are four or five "blue-hawks." "Chicken-
hawk" is another of the Keeper's names for
the bird. Here is an anecdote of a female
Sparrow-hawk which was hanging in the
larder, and now nearly weathered out of existence. The bird was found held fast by both
feet in a trap in which was also a young rabbit.
Seeing that the trap would take less than a
second to spring, the question arises, How did
both the creatures manage to get into it? The
Keeper's solution is that the rabbit was making
along an open "run;" and the hawk, seeing
it, skimmed along close to the ground, and
clutched the rabbit just as it struck the trap.

The Keeper's "red falcon" is the beautiful
Merlin, and of these there are a male and female
hanging in the larder. There is also the mottled
plumage of a Buzzard trapped during the snows
of winter; and by its side the silver-crested

head and wondrously adapted feet of an Osprey or Fish-hawk. This was shot among the rocks, in winter, down by the bay. There are a number of Carrion Crows, birds that destroy great quantities of the eggs of game. These, too, were shot in the side-channels when the weather was severe. They go there to feed upon floating offal and the various crustaceans that are washed up by the tide. Below the Crows are the gaudy remains of blue-winged Jays, and the glossy purple and white plumage of once-audacious Magpies. The grey-pated Daw is there, a single Hooded Crow, and the still more rare Great Grey Shrike.

This list betrays a deplorable want of discretion on the part of the old Keeper. He might well have spared the Kestrel, as it rarely flies at larger game than mice and cockchafers. In the vicinity of farmsteads and ricks this little falcon does almost incalculable good in destroying the small rodents that swarm in such places. The Sparrow-hawk is more difficult to defend ; few birds come amiss to it as prey, and at certain seasons it destroys quantities of young game. A full grown Partridge it can and does occasionally pull down. But the beautiful Merlin,

H

first among the falconer's favourites, with all
its ease and grace and majesty of wing—the
Merlin might be let live; it does but little harm,
feeding principally upon larks and pipits. Sometimes in autumn this roving arab of the air
destroys a few young grouse; but then by so
doing it confers a boon on the sportsman. These
are weak and ailing birds—the ones, probably,
in which are the germs of disease—and the moor
is best rid of them. The Buzzard occasionally
performs the same office; though, seeing that as
many as sixty mice have been taken from the
crop of a single bird, there is no justification
for its destruction.

In spring the garrulous blue Jays make free
right and warren of the peas sown in the old
man's vegetable garden; as well as the beans
which are grown in a clearing for the pheasants; and seeing that he sucks the eggs of
other birds, it is little wonder that the Keeper's
hand is against him. In addition to all this,
the Jay does indirect harm, which greatly
multiplies the cunning engines devised for his
destruction; for by pilfering the crops before
mentioned, which are planted with the object
of keeping wandering pheasants on the land,

a poor show of birds may be the result when October comes round, whereby the Keeper's reputation suffers. Even the audacious Pies will steal pheasant and partridge chicks; but the Shrike or Butcher-bird does little harm, as its shambles on the blackthorns prove.

One of the rails is wholly occupied by Owls— white Barn-Owls, brown Wood-Owls, and one of the Short-eared species. The first haunts ruined buildings, the second the dark and sombre depths of woods. There is little justification for any of the birds being here, except that now and again the Wood-Owl may perhaps snap up a young rabbit, a leveret, or a small bird. But even this is rare, and when developed is more an individual trait than a characteristic of the species. The Short-eared Owl occasionally kills young grouse on the moors or mosses where it breeds; but birds and insects are its usual prey.

Upon one occasion we were ferreting in a dark pine-wood, the floor of which was everywhere tunnelled with rabbit burrows. Into one of these a polecat ferret was turned, when there appeared hissing at its mouth an inflated bunch of feathers set off by a pair of

great, round, staring eyes. On closer examination, we found that a Brown Owl had made
her nest in the burrow; and although it was
December, and the ground snow-covered, the
nest contained an egg and two young ones.

A miscellaneous row on the vermin rails com-
prises Moles, Weasels, and Cats. The Mole is
libelled by being placed there ; he is a destroyer
of many creatures which are injurious to land.
Domestic cats soon revert to a semi-wild state
when once they take to the woods, and are terribly destructive in the coverts. They destroy
pheasants, partridges, leverets, and rabbits.
The life of these wild tabbies is wild indeed.
Every dormant instinct is aroused ; each movement becomes characteristically feline ; and
when these creatures revert to life in the woods
it is impossible to reclaim them. Climatic in-
fluences work remarkable changes upon the
fur, causing it to grow longer and thicker.
The cats take up their abode in some stony
crevice or hollow tree. In summer, when
kittens are produced, the destruction of game
is almost incredible. Of course the domestic
cat is quite a distinct species from the Wild
Cat of Britain.

The Keeper's indiscretions are fewer in fur than feather. His larder abounds in long-bodied creatures of the weasel-kind. There is the richly coloured, dark brown fur of the Pine-Marten; that of the Polecat, loose and light at the base but almost black at the extremity; and many skins of Weasel, reddish brown above with the sides and under parts white. For each of these creatures he has quaint provincial names of his own. The Pine Marten he calls " sweet-mart," in contradistinction to the Polecat, which is the " foumart," or " foul mart "—a name bestowed on the creature because it emits a secretion which has an abominable stench. There is also the Stoat or Ermine, which even with us is brown in summer, white in winter; the tip of the tail, however, is always black.

The beautiful Martens take up their abode in the rockiest parts of the woods where the pines grow thickly. They are strictly arboreal in their habits; and, seen among the shaggy pine-foliage, the rich yellow of their throats is sharply set off by the deep brown of the glossy fur. With us they do not make their nests · and produce their young in the pine-trees, but among the loose craggy rocks. Martens rarely show themselves till evening. They prey on rabbits, hares, partridges, pheasants, and small birds; and when we say that, like the rest of the *Mustelidæ*, they kill for a love of killing, it is not hard to understand why the Keeper's hand is against them. They do great harm in the coverts; the old man shoots them, traps them, and does them to death with various subtle engines of his own machination. To-day the Marten is rare; soon it will be altogether extinct.

Weasels do much less harm. They are the smallest of our carnivorous animals, and will probably long survive. They often abound where least suspected, in the cultivated as well as the wildest parts of the country. They take up their abode near farmhouses, in decayed outbuildings, hayricks, and disused quarries; and are often seen near old walls, or running along their tops with a mouse or bird in their mouths. These things form the staple of their food. There is no denying, however, that a Weasel will occasionally run down the strongest hare; and that rabbits, from their habit of rushing into their burrows, become an easy prey. But this does not often happen. To rats the Weasel is a deadly enemy; no number of them will attack it, and the largest singly has no chance against it. Like the Polecat, the Weasel hunts by scent. It climbs trees easily, and takes birds by stealth. The Keeper tells me that he has seen a brooding partridge taken in this manner; and on winter evenings the sparrows roosting in holes in hay-ricks. Weasels also kill toads and frogs; their mode of killing these, as well as of despatching birds, is by piercing the skull.

The Polecat, or Fitchet, keeps much to woods, and feeds mostly on rabbits and game. But in the northern fell districts, it often takes up a temporary abode on the moors, during the season that grouse are hatching. Then it not only kills sitting birds, but sucks the eggs, and thus whole broods are destroyed. Many "cheepers" of course fall victims. Knowing well the ferocity of the Polecat, I believe that the damage done to grouse-moors where this bloodthirsty creature takes up its abode can hardly be estimated. Like others of its tribe, the Polecat kills more prey than it needs. Sometimes it makes an epicurean

repast from the brain alone. Fowl-houses
suffer considerably from its visits; and it has
been known to kill and afterwards leave
untouched as many as sixteen large turkeys.
In the nest of a fitchet which was observed
to frequent the banks of a stream, no fewer than
eleven fine trout were found. The Gamekeeper
persistently dogs this creature both summer
and winter. In the latter season, every time
it ventures abroad it registers its progress
through the snow. It is then that the old man
is most active in its destruction. He tracks the
vermin to some fence or disused quarry, cuts
off the enemy's retreat, and then unearths him.
Trapped he is at all times.

The Stoat or Ermine is almost as destructive to
game as the animal just mentioned. Upon occasion it destroys great quantities of rats, though
this is its only redeeming quality. Partridges,
grouse, pheasants, all fall a prey to the Stoat, and
hares when pursued seem to become thoroughly
demoralized. Water is no obstacle to the
Ermine, and it climbs trees in search of squirrels,
birds, and eggs. A pair of Stoats took up their
abode in a well-stocked rabbit-warren. The
legitimate inmates were killed off wholesale,

and many were taken from the burrows with
the skulls empty. The Stoat progresses by a
series of short quick leaps, which enable it to
cover the ground more swiftly than could pos-
sibly be imagined of so small an animal.

Enough has been said to sketch the cha-
racters of these creatures, and to justify their
presence in the Keeper's larder. Interesting in
themselves, as wild denizens of the woods they
are fatal to game-preserving. But yet, with
what indignation did I lately see a gamekeeper put his heel into the nest of a Merlin
containing four bright rufous eggs!

Foxes abound in the fastnesses of the fells,
and the little wiry foxhounds that hunt the
mountains in winter account for but few in a
season; and so it devolves upon the Shepherds
and Gamekeepers to deal with them. This
they do irrespective of season. If allowed to
live, the foxes would destroy abundance of
lambs in spring. They are tracked through
the snow in winter, shot in summer, and destroyed wholesale when they bring their young
to the moors in autumn. It therefore happens,
that even the bright-red fur of the Fox may be
seen on the Keeper's gibbet. Hedgehogs are

taken in steel traps baited with a pheasant's or a hen's egg. At times Squirrels are killed in hundreds, but they do not grace the larder, as neither do the spiny Hedgepigs. Squirrels bark young trees, especially ash-stoles and holly.

Occasionally, a creature more rare than the rest adorns the larder. The old Keeper remembers a White-tailed Eagle and a Great Owl. Sometimes a Peregrine is shot, and more rarely, in autumn, a Hobby or a Goshawk.

CHAPTER IX.

THE fell Fox of the northern hills is an entirely different animal from the covert-fed animal of the south and midlands. It is of the "greyhound" variety, and endowed with great powers of speed and endurance. It is on this account that so few of the northern hill-country Foxes are killed in the legitimate fashion. Although the Cumbrian yeomen are keen sportsmen, the hill Foxes are hunted for reasons far other than those of sport pure and simple.

Vulpecide is not even a crime, a price being set upon the head of the Fox by the wardens of every mountain church. And this, where the crags afford such harbour, and the holdings are essentially sheep-farms, must necessarily be done. With no game-preserving, and but

little life on the mountains, the foxes prove
very destructive in winter. When other
supplies fail, and reynard is denied the barn-
yard, he has recourse to the small black-faced
sheep. These, from their small size, are not
difficult to overcome, especially if dog and
vixen hunt in company; and it is evident, from
their shambles, that they find mountain mutton
very toothsome. It is this *penchant* that makes
the hill farmer so persistent an enemy of the
Fox. He traps it, shoots it, and when he can
safely do so, lays poison in its paths.

One of my amusements in long-gone school
vacations was to lie upon a green ledge of the
crags armed with an ancient flintlock, and from
this point of vantage deal death to the cubs as
they came out to play at the mouth of their
den. The destruction of the young is the
farmer's method of keeping the species within
bounds, and is the only practicable one.

A few breeding sites supply a wide tract of
country; and these, for the most part, are in
quite inaccessible fastnesses. There the cubs
stay through the summer until early autumn,
being catered for in the meantime in the most
assiduous manner. The mouth of their earth is

a perfect shambles, where every species of native game is represented. Then there comes a time when the playful family is taken by night to the woods or the moor, and here, as harbour offers, they abide till winter. Colonies may not unfrequently be found among the heather; their vicinity being marked by the heads and wings of Grouse, Curlew, Plover, Dotterel, and Hares. Such a spot is a very paradise of fox-dom, and a perfect training-ground for the cubs. They gambol about at twilight, sending up clouds of fur and feather, evidently quite un-aware that this is the most critical period of their lives. When found under such circum-stances, a stout stick and a couple of dogs soon enable the shepherd to despatch a whole litter; and he never loses the opportunity. At such times the parents keep at a respectful distance, never attempting to defend their young.

The extreme beauty of the red mountain-Fox is best seen when his figure is sharply outlined against the snow. How fleetly buoyant he glides along, his brush floating light as air behind him! What grace in his leaps, and litheness in his long, finely drawn limbs! They are perfect masses of muscle. After look-

ing at them it is easy to account for his marvellous powers of endurance. That easy jog-
trot, by reason of its very buoyancy, conveys
but little notion of the actual speed. As the
Fox floats along, his brush lends itself to the
delusion, the close thick fur concealing the intense muscular play beneath. His speed can
only be judged by comparison.

One day in winter, as I listened to the baying
of the hounds among the crags, a fine dog-fox
leaped lightly over the fence within a few feet
of where I stood. A fast cur bitch ran him
through a long meadow, but only kept pace for
a few strides, the Fox out-distancing her with
contemptuous ease. The race opened my eyes,
for the bitch ran like a rough-coated greyhound, and was "Fleet" both by name and
nature.

Like many wild creatures, Foxes have well-
defined routes, seeking their food morning and
evening. A curious fact, and one perhaps
hitherto unrecorded, is that, when foraging,
reynard's whereabouts is often made apparent
by carrion-crows and other birds that noisily
pursue him high overhead until he secretes
himself. In mountain districts his fare is often

meagre and hard to find; and the shifts he makes and the pittance he has to put up with must sorely pinch him in winter. In summer, mountain hares and wildfowl are not difficult to obtain; but in times of severity these creatures descend the mountains, drawing nearer to the haunts of men. Then he is glad to get rats and mice, even beetles and earthworms. Hedgehogs' skins may be found near his "earth," and when frogs are obtainable he considers them delicacies. Upon one occasion I found stored up twenty-three Shrew-mice; though why they should have been stored instead of eaten I cannot conjecture. Strangest of all, Foxes are extremely fond of fresh-water Crayfish, obtaining them from the mountain streams in summer when the water is low.

A curious habit, and one in which the mountain-Fox invariably indulges, is that of frequently stopping to listen when leaving the " earth." At first about a hundred yards divides each halt, but when further away the distance increases. In retiring to the crags, foxes never enter from below—always from above. Owing to its tread being much lighter than a man's, and its hearing quicker, it is rarely

that even dwellers near its haunts get a peep
at him. If surprised, he is never disconcerted,
but trots off with the most unconcerned air
conceivable. One day, walking by a fence
which skirted a fir-plantation and suddenly
rounding a curve, I observed a Fox coming
towards me; when we mutually stopped to
gaze at each other. On the part of the Fox the
hesitation did not last long. In a moment he
again came leisurely on. When a few paces
in front, however, he took the fence at a
bound, kept close beneath its further side for
some distance, and then, slightly exerting himself, was soon out of sight.

To show the stamina of mountain Foxes, of
the hounds which are bred to hunt them, and
of the yeomen hunters who follow on foot,
one remarkable run may be cited by way
of illustration. This lasted upwards of nine
hours, and the distance covered must have been
considerably over a hundred miles. The chase
began about noon, and at six in the evening,
when reynard was believed to be exhausted,
he again made for the hills, where both fox and
hounds were lost to the hunters. At nine the
hounds were heard returning by the way they

had gone, and were still in full cry. By this time half the pack had fallen off, and the echoes that rang among the mountains in the moonlit night, as the hounds passed and repassed through the gorges, were magnificent. Soon the prolonged deep baying was changed into short sharp barks—a sure indication that the dogs were viewing their game. In a short time all was still, and then, perhaps, ended the life of the toughest old fox that ever ran the fells. The hounds returned, showing by their torn faces that reynard, even when run down, had fought desperately. Many of the pack, however, were lost or exhausted, and did not return until next day ; one, completely worn out, crawled from the hills after a week, and three were found to be "crag-fast." Rescuing these is always a dangerous business, and on this occasion it was performed by a party of shepherds with the aid of ropes. I have set down the above story as it is told by the shepherds, though certain of the facts seem almost incredible.

Here is another actual incident of mountain fox-hunting. A man named Dixon fell from an overhanging precipice three hundred feet in

I

height; and, although terribly bruised, and
almost scalped, he broke no bones, and recovered
from the shock. In falling, he struck against
the rocks several times; but the story goes that,
on coming to the ground, he sprang at once to
his knees, and cried, " Lads, t' fox is gane oot
at t'hee end ; lig t' dogs on "—then fell down
insensible. The place has since borne the name
of " Dixon's Three Jumps."

Of all pastimes of the northern yeomen, that
of hunting is the most popular. Most of the
ballads sung upon festive occasions have for
their subject-matter some " Bet Bouncer," with
cheeks " broad and red as a parson's cushion,"
or memorable fox-chase. Wordsworth asserted
that in his time, when the hounds were
out on Helvellyn, not a soul in the village
would stay for want of leisure to enjoy the
sport. And I know of my own frequent knowledge that this applies to the daleswomen as
well as the men. Hunting with Harriers or
Fox-hounds is equally popular, though from the
nature of the ground the former has most female
followers. So strongly did this love of sport
exist among the yeomen of the last century,
that it is even asserted they went hunting

on Sundays. In time this sabbath-breaking and lawlessness became so flagrant that the parson from the pulpit openly denounced it. And it is said to this day that his denunciation was uttered in this wise: "O ye wicked of W——, if you go a-hunting any more on the sabbath day, I'll go with you!" and the good clergyman, being as fond of hunting as the rest, kept his word to the letter.

CHAPTER X.

THE HAUNT OF THE ANCHORITE.

WE are lying on the confines of a wood where we have lain for an hour. Such a multitude of sights and sounds are around that the eye and ear seem hardly capable of absorbing them. And yet there are no blurred impressions, no confusion of sounds. The eye faithfully produces each picture, the ear each soft swish of the pines. Is the beauty of this leafy woodland way in itself, or only in the eye of the observer ?

> Are these sweet sounds of the early season,
>   And these fair sights of its early days ?
> Are they only sweet when we fondly listen,
>   And only fair when we fondly gaze ?

The poets have told us that what we call nature is but our own conceit of what we see ; and doubtless they are right. We find our own complexion everywhere, and receive but what we give. And so it becomes literally true that

> There is no glory in star or blossom,
> Till looked upon by a loving eye;
> There is no fragrance in April breezes,
> Till breathed with joy as they wander by.

Doubtless it is this unconscious appeal to the inner self that evokes the sense of reverence which we feel when standing face to face with nature's finest objects. We interpret ourselves rather than the things about us. These are tangible, and we idealize and create them afresh.

Under floods of beechen green and shadows numberless the Warblers are singing of summer in full-throated ease. These are in the deepest recesses of the wood, and the sounds only faintly, and at intervals reach us. Quite a wealth of woodland beauty is around. The leaves of the grey-boled Beech are of the most delicate green, as are the long trailing tassels of the Pine. Soft mosses cover the floor of the wood. A small green warbler restlessly flits among the tangled weeds. It complains in melancholy " tweet, tweet," that we have in-

vaded its haunt, and are near its nest. The
warbler and its kind we have come to woo—to
pry into their secrets. Show the bird that you
partake of its nature, and it will trust you.
The most shy and retiring ones soon do this.
And so we watch and wait and are patient.

Wood-flowers are all about us. These are
stunted, and their colours subdued; the sun
only quivers through the beechen boughs in
frescoes. The slender pines better let down
the lines of light. Beneath them the flowers
respond; they seek to kiss the light, and shoot
upward. Where we lie the flowers are of
spring; those under the pines are of summer.
The colours of these are bleached; those are
intensified. Here are hyacinths, anemones,
and wood-sorrel; there foxgloves, woodbine,
bellflowers. As Summer advances she deepens
her train. Follow her, and she goes from
green to gold, from gold to russet. Only
the birds that have business seem to be in
the wood. Except the sounds of our wood-
bird's " cheep," everything seems afar off. In
the deepest recesses of the wood there is little
food, but outside yonder myriads of gauze
wings disport in the sunlight. Still we are

waiting and watching. Presently the green-brown bird drops down, and the flowers which we have been admiring hang over its hidden nest. It is a common species, and we pass on.

The sombre and twilight of dark woods is pleasant at times; but for a moment, to be lifted out of it, we span a long straight pine and climb by its boughs to the top. Looking out from our aërial altitude, the world seems a flood of delicate greenery. No tree so beautiful as the Pine. A thousand tender tops seek seek the light—the only moving objects in the landscape. Trailing green-tasselled fingers, how exquisitely beautiful ye are! How delicate your tracery! And then the balm and gum of resinous woods! In summer the long pine-boughs sit like brooding doves over the warm earth. In winter they hang out funereal plumes when the ground is locked in ice. No dead crackling boughs are here—nothing but life; the warm yielding up, the hum and essence of being. Among the fir-tree tops all is sunlight. Squirrels chatter, wood-pigeons coo. Even the flies have come up here, and lazily revolve in their mazy flight.

Peeping out of the wood, far out yonder, is

a Ruin—a ruin with all its monastic associations
and hoar. When the red deer and the wolf
roamed along the fell-side, it formed a Hospice.
Later it was the meagre shelter of an Anchorite
—a Recluse. In it he lived, harmless and un-
offending. He knew well the times and seasons, gathered flowers and boughs and berries
about his lone home, and talked with the
animals and birds. The latter he fed daily;
the former learned to confide in him. He
dug ground-nuts for the badgers; he hoarded
for the mice and squirrels; rabbits rustled
away, though not terrified. Near his haunts
they found green corn-stalks. Hares looked
from the green brackens and skipped in the
moonlight. Even the bead-eyed weasels lived
with him among the stones. Only the fox
never came near.

The stones were old and hoar and lichened;
the same might have been said of him. He
loved everything, and was loved by all. But,
like the Ruin, he too passed away. When the
berries of the Deadly Nightshade came and the
leaves fell, he seared and died. Under that
mound he is still a Recluse—the Spirit of the
Wood. But this is a strange dead thing. We

are of the sun; we are of the light. We live,
and move, and have our being; we cry for air
and sunshine. Away with death—that drear,
sombre, silent thing.

Still stand we by the Ruin. A bit of sunlight comes upon the dead stones; it is a bird.
a bright, lifeful, joyous creature.

CHAPTER XI.

MOUNTAIN SHEEP.

SCATTERED along the slopes of many of the northern valleys, there still lingers the last remnant of the old yeomen, or 'Statesmen class. Their houses are strongly built of stone, and are essentially those of a utilitarian age. Each homestead has about it a few fertile fields —meadows which margin the valley stream. These are sufficient to afford keep for a dozen milch cows, and in summer yield abundant crops of hay. The young cattle graze the " grassing heads " in summer, but are brought to the coppice belts of birch and hazel to pick a scanty winter fare. There is no ploughing, and therefore few horses are required. Although the 'Statesman, with all his virtues, is rapidly becoming extinct, neither political nor agricultural economy can alter nature's decree that these small holdings must ever remain sheep farms. Each farm in the dale has its "Lot," or Allotment, on the fell, which feeds from five hundred to a thousand sheep. This number is about the normal one, though some of the largest farms have most extensive "heafs," and graze from two to five thousand sheep. These are of the Black-faced, Scotch, and Herdwick breeds. All have coarse, hair-like wool; the Scotch and Black-faced have horns, whilst the Herdwick is polled. Yet each wears what the hill farmer terms a "jacket and waistcoat"—that is, long wool without, with a soft, thick coating beneath. This is the one great characteristic which fits the animal for its life among the mists. All the breeds indicated are small boned, and produce the best and sweetest mutton. It is the tending of these that constitutes the chief work of the dalesmen throughout the year.

We have said that each farm of the valley has allotted to it its hundreds or thousands of acres upon the fells, and it is wonderful how the sheep know their own ground. Of course

this was the more remarkable before the enclosure of the Commons, when only a stream,
a ridge of rock, or a heather brae formed a
nominal boundary. Now hundreds of miles of
wire fence stretches its dividing influence over
the wild fells, and is the means of destroying
great numbers of grouse. One of the provisions for localizing sheep upon their own
" Lot " is as follows. When a retiring tenant
is leaving his farm, he is allowed to sell or
take with him, say, three-fourths of his flock
of two thousand sheep; but the remaining five
hundred must be left on the old ground. It
is imperative upon the retiring farmer that this
nucleus be left, though sometimes the whole
flock is taken by the incoming tenant, and so
remains. In any case he must purchase the
number to be left upon the " heaf " at a valu-
ation by one of the dalesmen or respectable
yeomen, mutually agreed upon by the landlord
and himself.

In each parish there still exists at some
farm a *Shepherds' Guide,* setting forth the tar
marks, smits, and ear-slits, peculiar to the
sheep of each farm in the township. This
book is in the keeping of some responsible

person, and is used as a reference-book in cases
of dispute. It sets forth the name of each
farm, the number of its heaf-going sheep, a
rough definition of their range; and finally,
the account of each flock is illustrated by cuts.
This shows, to take an example, " I. B." on
the near shoulder, a red smit down the flank,
with the near ear slit down the middle. The
" smits and slits " are essential, for, although
the initials of the owner may, and frequently
do, become blurred and indistinct, the former
are lasting, and, in case the animals have
strayed, they may at once be identified. With
the enclosure of the Commons, this "Smit-Book "
is now rarely used, and no recent edition has
been printed.

Most of the sheep winter on the fells. On
the highest of these, in severe weather, they
have usually to be foddered through three or
four months of the year. Hay is taken in peat
" sleds," and bundles are thrown down at
intervals. Failing this, the sheep are expert
in scraping away the snow to get at the buried
herbage. This they do with their feet and
noses, and, as they clear away the snow,
the grouse (though this applies only to the

lower ranges) follow and eat the heather seeds from beneath the bushes.

Sometimes a whole flock of sheep is buried deep, and has to be dug out. Even taking it for granted that the whereabouts of the entombed flock is known, the task of rescuing them is one of great difficulty. In attempting it, the shepherds have occasionally lost their lives. The animal heat given off by the sheep thus buried thaws a portion of the snow about them. Stretching their necks over this limited area, they devour every blade of green, even the turf itself. This exhausted, they eat the wool from each other's backs. Under these circumstances, the tenacity of life shown by the sheep is marvellous, and many have been rescued alive after being buried for twenty-eight days. When brought to light, these poor creatures are in a weak and emaciated condition. During the long and terrible winter of 1885, the fell-sheep suffered severely. On the higher runs they perished by hundreds. The farmers (four in number) of the farms lying contiguous to Sca-Fell, alone lost fifteen hundred sheep out of an aggregate of about six thousand. The whitened bones and fleeces of these were dotted everywhere about the fells, and to the hill farmers, in these times of depression, this fact almost spelled ruin. The skeletons were bleached, and the only things that profited by the protracted snows were the peregrines and ravens of the crags. These birds still find an asylum in the deepest recesses of the mountains.

In the desolate hill tracts winter usually lasts through eight months of the year. Layer upon layer of snow become hard frozen, and upon the highest peaks of Skiddaw and Sca-Fell this often lies till June and July. During Midsummer Day of 1886, the mountains were all day lashed in blinding snow. For the most part, April clears the summits of the mists, and a better time is at hand. The snows have gone from the lower grounds, and sparse vegetation comes sweet and green. This grows quickly, and the flocks rapidly gain in condition. Now the sheep are ever active; by the torrent sides, by the leas of the boulders, along the rock-ledges they seek the freshest grass. And in search of this they sometimes become "crag-fast"—that is, they climb and climb from one narrow ledge to another, sometimes placing their four feet even

upon a jagged splinter. If a face of rock
intervene, and they cannot climb out to the
top of the crag, they turn to descend. But
here retreat is cut off. Sometimes they remain
in this position for days, eating whatever is
within reach, when one of two things happens.
Either they are rescued by the shepherds, who
are let down to them in ropes; or they fall
a prey to birds and foxes. The raven, the
peregrine, and the buzzard freely appreciate
the creatures' position, and await their chance.
Sometimes the birds so terrify the sheep, that in
its fright it makes one mad leap, and is dashed
to pieces as it descends the crag. Then the
raven hardly waits until death has come, but
immediately goes dallying round and round the
carcase, and soon falls to work upon brain, lip,
or palate. The peregrine feeds only so long as
the flesh is sweet, though the hill foxes and
carrion crows visit the spot for a week.

Snow-lines are sketched along the stone
fences of the fells; but this is all that remains.
Everything testifies to the coming of spring.
The foaming fell-becks sparkle in the sun,
and the climbing sheep are sprinkled over
the crags. A breadth of blue is overhead, and

towards this they always climb. When the weather is fine their heads are infallibly turned towards the skyline.

And now the shepherds are busy with their flocks. The ewes are drafted out and quietly driven to the lowlands. These are distributed among the fields of the hill-farms, and for a time have better fare. An anxious time is approaching; but here the lambing season comes fully two months later than in the lower cultivated valleys. Daily attention is paid to the ewes, and about mid-April the lambs begin to make their appearance. The Black-faced and Herdwicks are hardy; there is no folding, no extra feeding, and they come through the critical time in a manner that would astonish southern farmers. The mortality is exceedingly small; the lambs are strong and quickly on their legs. As soon as the lambing season is over, and the little strangers are strong enough to bear the journey, the whole flock is driven back to the fells. Each year the farmer breeds two varieties of lambs. The Black-faced and Herdwick ewes produce both, one of which is half-breed, the other pure. The pure portion is to keep up the blood of the

K

farm; the half-breeds, which are heavier and larger lambs, are intended for sale. At this time the barren ewes are also drafted from the flock, they too being fatted for the market.

As the warm days of May pass to those of June, the shepherds commence to "gather" their flocks for the washing. In this they are aided by collies—small wiry creatures, almost inconceivably intelligent. They in nowise resemble the sheep-dogs of the show-bench, but are mostly built on the lines of the hill-fox. They can be hounded for miles—as far as they can see the action of the shepherd directing them. In fact, they are quite knowing enough to work without this direction; and I have seen them scale a crag and carefully bring a flock of sheep from the rocks and ghylls where not a single living thing was apparent to the eye.

"Devil's "Dust," Wily," and "Fleet," were three of the most intelligent brutes that ever ran. I have spent weeks among the mists with this lovable trio. When a headstrong Herdwick gets upon the shelving rocks of the crags, the dogs never force. They crouch, using the utmost patience, and rather guide the sheep than drive it. That these dogs become fascinated with their work there can be no question. It is clear, too, that it is difficult, and always more or less painful; for after a hard day's running upon the fells their feet are dreadfully cut by sharp stones, which in summer blister the hand if laid upon them. The beds of flat tinkling stones which do the damage, give out their not unmusical notes as men, sheep, and dogs rush over them. It is usual on the hill farms, where a great number of sheep are kept, to work the collies in relays. A brace are taken out one day, and rest the next. But at times of gathering for washing, or shearing, this plan is not always practicable, and all the dogs are working at once.

I have said that it is at the time of gathering the sheep for washing and shearing that the dogs are hardest worked. When a fine spring has reduced the fell " becks," and the clear water lies deep in the pools, then it is that washing takes place. The water is now tepid ; and by the side of the deepest pool a bit of bright turf is encircled by wooden hurdles, and a fold constructed. The shepherds have been out on the fells through the short summer night, and now down the corries long lines of sheep

are seen approaching, all converging on the
rugged mountain road. The sheep and shepherds are met by a group of fell folk who have
come to assist. These are the 'Statesmen and
their sons, dalesmen from the next valley,
neighbouring herds, and often some women.
Sorting the sheep and depriving them of their
lambs is gone through, the scene meanwhile
being most animated—men shouting, dogs bark-
ing, sheep stamping and fighting the dogs,
while others lightly top the hurdles and attempt
to make back to the fells.

Two strapping yeoman wade into the water
to their middle, and the business of the day
commences. The washing of six hundred sheep
means a long summer day's work; and now
all exert themselves to the utmost. Two men
take each sheep with both hands and heave
it into the pool. Here it is caught by the
washers, well soused, then allowed to swim
to the opposite bank, where for a moment it
stands dripping, then moves off to the sunny
sward. Weighted with water, the creature is
stunned for a while, but soon begins to nibble
the short herbage. During the whole of this
time a constant bleating is kept up between the

lambs and their dams; nor does it cease until they are brought together after the washing to be driven back to the fells. By this time every one engaged in the day's work has imbibed much strong ale; but hard work has rendered them none the worse for their deep draughts. Seeing the sheep sprinkled over the fells a few days after this, their coats are observed to be whiter and the wool more fleecy.

Washing, of course, is preparatory to shearing; and this comes a fortnight later. All the dale responds. Goodwill is one of the characteristics of the 'Statesman. For shearing, as for washing, the sheep have to be gathered; and this sometimes takes two days and a night to accomplish. The animals are brought down the mountain road to the farm and placed in rude stone folds, each holding perhaps a hundred sheep. The Shearers arrive from up and down dale, and among them come the Parson and 'Squire, each in white "overalls." The Shearers seat themselves on "creels" ranged round the main fold, and a dozen stout lads act as "catchers" to supply their elders with sheep. Bright bands are produced to tie the

goatlike legs of the Herdwicks ; and the flash
and metallic " click " of the shears are seen and
heard afar.

Soon the scene is one of picturesque animation. A turf fire is lighted, and upon this
a pan of tar bubbles and boils. Standing by
it are the owner of the flock and the Parson.
They stamp the former's initials and smit marks
upon the sleek sheep just freed from their
cumbersome coats. The 'Squire goes round
among the Shearers and acts as " doctor." He
carries a small can containing salve and tar,
which he applies to the cuts accidentally made
in shearing. These snips are common, but arise
more frequently from the sheep's kicking than
from carelessness in the shearer. Two or three
girls rid the clippers of the stripped fleeces ;
and the latter, together with the fallen wool,
are placed upon the unhinged barn-doors by
two portly dames from down dale ; and are
then stowed away in the wool-loft. All the
flock is stripped. Then comes the banquet.
And such a one ! Huge rounds of beef, legs
of veal and mutton, quarters of lamb, hams,
and pies of every description ; there are sweet
puddings, and all things else in keeping. Then

the company withdraw to the barn, where the
creels are ranged round against the hay-mows,
and strong ale and trays of tobacco circle
among the guests. A long table runs down
the middle ; the Parson presiding at one end,
the 'Squire at the other. Glasses are filled,
smoke-wreaths begin to ascend, and the ballads
of the dalesfolk are sung. On such occasions
the Parson drank, sang, and smoked in as
orthodox fashion as the rest. This is said in
nowise disrespectfully. The Parson was one
of us, tilled his glebe, and had a sheep-run
on the fells. These constituted parts of his
" living."

At the shearing, the lambs are separated
from their dams, and receive the impress of
their master's initials as well as smits and
ear-splits. The half-breeds for the production of mutton are weaned from their ewes,
and are not allowed to return to the fells.
They are kept until autumn, sold at the great
northern sheep fairs, and are then sent to be
fatted on southern grass lands. Here they
feed quickly and make excellent mutton. Only
the pure-breed lambs—Black-face or Herdwick
—the future heaf-going sheep of the home farm,

are retained. After the "clipping," and whilst
the yeomen are carousing in the old barn, the
shepherds start on the return journey with the
fleeceless flock. As the lambs are brought
to the ewes there is a perfect babel of bleats.
Turned into the long lanes, the white fleeceless
flock presents an indescribable picture of pastoral beauty. Everywhere sheep hang upon the
hazel-clad slopes, stretching their quiet necks
to the herbage. Not a foot of the bank is
unoccupied—two long lines of sleek, browsing
sheep, reaching away until a bend in the road
hides them. Soon the bleating becomes less
general; then it ceases, and a strange stillness
fills the lane. A breeze brings up the left
lambs' voices, and all is confusion again. And
thus we plod slowly on to the fells in the sultry
summer afternoon, and turn the flock again
upon the green slopes. The hills become
animate with a thousand sheep. Soon few
are to be seen; they have dispersed, but seem
to have dissolved.

Then we turn homewards, ourselves and the
three dogs—not down the long dale road, but
by the Forest—"forest" only by name now,
and thick with peat, having traces of birch

and mountain ash. Our way lies along the "grassing-heads" running parallel to the valley, but high above it. Coming through these, rushes prevail, and hidden springs. Among the leaves the gad-flies rest, and grasshoppers make harmony with the hidden water. Then we come into scrub of oak, birch, and hazel. Flies abound, and a few birds.

From what has been said of the farms of the fell dales, it will be seen that wool is one of the chief products of the 'Statesmen. Among the many quaint buildings of the hill folds, one is usually set apart as the wool-loft. And it is deplorable to have to record that many of these, even the teeming barns themselves, are full of wool, the produce of many seasons' "clips." For the hill-farmer has felt depression in trade as well as his southern neighbour, though in a different way. Some of the yeomen say that they have four, five, even six years' wool harvests in their barns, and cannot sell it at present prices.

We have said that time was when the wives and daughters of the 'Statesmen spun the wool, and wove it into cloth. This was done in almost every house, and by this light labour the

long winter evenings were pleasantly beguiled.
It is somewhat strange that this occupation
was one much indulged in by the poorer clergy
who guided the spiritual lives of the yeomen.
Of one of these as a type we shall speak. He
assisted his neighbours at haytime and shearing, though instead of receiving money he
was paid "in kind." He also made wills,
butter-prints, and was Notary Public to the
whole parish. For these little offices he invariably chose wool as his reward ; and
for a reason. The tributary fleeces he was
wont to collect by the aid of a shaggy white
Galloway, with which he always tramped the
fells. Across the back of the old horse were
two panniers crosswise, in which the fleeces
were carried. The annals of his quiet neigh-
bourhood tell how for eight hours each day
he was occupied in teaching the children,
his seat being within the communion rails.
While they repeated their lessons by his side,
he was busily engaged with his spinning-wheel.
Every evening, too, he continued the same
labour, exchanging by way of variety the small
wheel at which he sat, for the large one on
which wool is spun, the spinner stepping to

and fro. Thus the spinning and winding filled up the interludes of his evening labour. The elder of his children assisted in teasing and spinning the wool, and at the whole trade it was well known that both he and his family were proficients. When the various processes were completed, and the whole ready for sale, the good man would lay it on his back, by sixteen pounds or thirty-two pounds weight, and carry it on foot to market seven or eight miles, even in the depth of winter.

And yet this primitive Parson was a pronounced type of his contemporaries. In his life he held three " livings," occupying the last sixty-six years. He died aged ninety-three.

During the time he was busily employed he never once neglected his more important spiritual duties. These he discharged zealously and faithfully; brought up, educated, and established in life a large family, and died "universally lamented." His fortune at his death, amassed by great industry, amounted to £2000, besides a large quantity of linen and woollen cloth spun by himself, chiefly within those communion rails of which we have spoken. The following extract is from a letter describing him " At Home" : " I found him sitting at the head of a large square table, such as is commonly used in this country by the lower class of people, dressed in a coarse blue frock, trimmed with black horn buttons, a checked shirt, a leathern strap about his neck for a stock, a coarse apron, and a pair of great wooden-soled shoes, plated with iron to preserve them—what we call clogs in these parts—with a child upon his knee, eating his breakfast."

Here is an interesting bit of conservatism which deserves some passing record ere it ceases to be. The Shepherds' Meeting itself is one of the oldest institutions of the hill districts, and is annually held at the little wayside inn on the summit of Kirkstone Pass. The period of its importance dates back to a time prior to the enclosure of the Commons, when it formed part of the economy of sheep farming in the North. It has been already stated that each dale farmer has right of " heaf" for so many sheep upon the Common; these being without any barrier to restrict their range. Yet it is remarkable how each flock knows and keeps its own ground; but the reason for this has been given. In spite of these shrewd precautions, however, a few

"strays" may always be found on distant
"heafs," and the hill shepherds meet annually
and reclaim them. The meeting is held at the
end of autumn, and for some weeks previous a
sharp eye is kept for intruders. On a certain
day these are driven to the rude stone folds
on Kirkstone, and every shepherd takes back
his own. He is greatly aided in identification by the institution mentioned—the "Smit
Book." Now that the Commons are enclosed,
this book is rarely used; and the decay of the
Shepherds' Meeting is to be referred to the same
cause.

Going back to the time of its importance,
the day was always interesting. Sheep, shepherds, dogs, and dalesfolk together made up
a scene of picturesque animation. After the
more serious business of the day, there was
much northern merry-making. Wrestling,
racing, leaping, and fox-hunting on foot were
among the chief diversions. Two neighbouring
'Squires sent each a pack of hounds, and then
entertained the hunters at dinner upon their
return from the mountains.

A mountain hare sometimes makes a good
run; and, when foxes are scarce, offers a capital

substitute. The writer well remembers a memorable hunt many years ago on a glorious day—the roads ice-bound, and the mountains snow-covered. After partaking of ale at the inn—one of the numerous " highest inhabited houses in England "—the whole assembled company set out, and at once ascended into a region of snow. There was a sprinkling of women and girls. From the elevated region which was quickly attained a glorious prospect soon presented itself. Below in the vale was the village of Grasmere, with all its associations. Upon the opposite steeps Silver How and Helme Crag, where lay the " lion and the lamb." Easdale and its silent tarn dipped between the two heights, and Dunmail Raise between Steel Fell and Seat Sandal. The sun shone upon the snow-clad hills, throwing up an exquisite rosy flush ; and this being reflected from the valley below, all nature seemed bathed in golden light.

Here is the ridge of Helvellyn, and we are above the snow-line. The hounds have previously thrown off in a plantation at Fairfield's base, and the huntsman has to use all his skill to prevent the pack from dividing, and going off after two hares. And now ensues one of the longest runs of the season. Right along the breast of Fairfield—no shifts nor doubles, but straight ahead after the manner of a fox. The hare leads over Fairfield, along the steeps of Nab Scar, to Rydal Head, down into Rydal Vale. The quick ascent over the snow-clad hills from the neighbouring valley has put to test the stamina of the straggling followers, and three only reach Nab Scar. Here, as the hounds are bringing puss back from beneath, we stop to look about us. Away to the south stretches Windermere, with its brown slopes and shaggy copse; Rydal and Grasmere lie below, while over yonder is Esthwaite and Coniston Water; then Easdale, Elterwater, and Blenham Tarns. Far to the south, like a flood of molten fire, stretches Morcambe Bay, its boundary hills rolled and lost in cloud.

Soon the hounds recross the shaggy top, and return in one long line, presenting a pretty picture against the snow. As they take us back along Fairfield, many of them are visibly fagged, and the line becomes long and straggling. Here a raven, roused, flies off croaking to the nearest glen. The tracks of the hare

and hounds are sharply marked against the
snow, and something may be read from them.
Here, puss has run in a circle, then made a
long leap at right angles to her course; but
all in vain. The hounds are baffled only for a
second, and soon take up the quest. Then the
descent begins, and the hare makes back to the
woods; and here, to our delight, the dogs are
whipped off, every one agreeing that such a
hare is worthy of life.

CHAPTER XII.

ON THE MOORS.

BY A SPORTSMAN'S WIFE.

WE went North about the beginning of August,
and the " grouse fever " set in on the tenth.
Once in the previous night I had heard my husband shout excitedly in his sleep " Mark over! "
but when I awoke him he only asked, " What's
the weather?" and recommenced to snore. In
the morning he was irritable, and constantly
looked out of doors. Each time he returned
he told us that the mist caps still stuck to
the hills, that Ben Bald was hidden in clouds.
Like the Wandering Jew, he ever and ceaselessly paced from room to room of that Highland Shooting-box, and asked querulously of
every one who came near him, " Will Ronald
never come?" None knew whether " Ronald"
was dog or man; but we soon found out.

L

At last he came—a young Highlander; and, as I want what is here set down to be veracious, I must state that he was indecently clad. He wore a kilt; his legs were dirty, as was everything about him. He had a huge pair of sandy whiskers, and I should think that he never either washed himself or combed his hair. I suppose that when nature decreed, his garments sloughed off, as does the outer cuticle of a snake. But be this as it may, he acted as soothing balm to the perturbed spirit of my husband. I put it on record that upon the arrival of the excrescence referred to, the amiable being to whom I am wedded came near driving the servants dead with fright, by coming to the kitchen and rushing out with a kettle containing boiling water. It was the first and only act of menial labour of a lifetime. They thought he was mad. I heard screams from the scullery, and followed to the dining-room. He was brewing whisky—"for Ronald." Ronald took snuff, Ronald smoked huge black coils of something which they told me were slow-worms. It was a curious custom they had in the Highlands; they all did it.

My husband had taken a Grouse Moor and

a Deer Forest in one. He was not to kill "more than fifteen hundred Grouse or twenty Stags." That stipulation I myself saw, and the agreement was solemnly signed and sealed. Ronald had come to report progress as to the past breeding season. They adjourned to the gun-room, leaving a strong scent of punch all along the line. As my husband could not speak a word of Gaelic, and Ronald knew but little of English, I suppose the report received was satisfactory. Our little Jack, who had dared to invade the gun-room, told us afterwards that they did "like this"—and he simply waved his arms about and made faces. At dusk the gillie went away.

Next morning was the Twelfth, and we were up betimes. Although none of us had ever seen the sun rise, we all saw it then. It came from behind the hills, and spread an exquisite rosy flush over the landscape. The white mists rolled up and over the mountains, and some strange sounds which we heard, Ronald said were the black-cocks crowing. A beautiful lake lay shimmering in a green hollow beneath, and I must admit that we were set down in a wild kind of paradise. Strange mountain

flowers were blooming about us; the birds were
different to those we had left behind; and there
was something in the air indescribably sweet.
The rains and mists which had prevailed since
our arrival were gone, and the mountain becks
tore down the slopes, transformed to torrents of
silvery foam. It was the first time we had
seen about us, though that one look was the
best part of the " shooting."

At nine, " four guns " (so they called themselves) started up the " corrie " with as many
dogs. I believe they said they were going to
" shoot over " these latter, and not at driven
birds. This was the old-fashioned method, we
were told, and the most sportsmanlike. The
quantity of " Glenlivet " which those people took
with them would have formed a fell " beck " in
itself.

At evening they returned, bringing with
them blue Hares, Ptarmigan, some Black-game
(which were not then in season, and were
shot by my husband), forty brace of Grouse,
with a few Plover and Snipe. The thirteenth
was an " off day," and we all went out to
gather heather. In it we packed the grouse,
which were to be sent south. Upon this occa-

sion I managed to stick fast in a peat-bog, but did not much mind, as I saw for the first time the béautiful Grass of Parnassus and some rare wild orchids. We tried to climb a mountain, but found climbing terribly arduous, and had to return home. Mountains, with their filmy blue hazes and wondrous purple lights, are exquisite afar off; but near, we found them less pleasing.

The fourteenth was another shooting day, the bag being as before, with the addition of a Roe-deer. Subsequently we tried the venison of this, but found it unpalatable. Whether the fault lay in the cooking, I don't know; but it certainly was not in our appetites.

Three days of deer-stalking succeeded; the bag at the end of that time being a sprained ankle, and the sight of the antlers of a Red Deer about two miles off. Disgust I saw had long been brewing; now it was openly expressed. Several attempts at driving the deer had been made, but the monarchs of the glen always scented the hunters from afar, and went the wrong way.

On the last night of our sojourn in the Highlands (so it turned out) there was an air of mystery about " the guns." There was some-

thing about "an express bullet," "slippery
rocks," "glancing off," but more than that we
could not gather. I have my suspicions what
was meant. Next day we went south. There
was one little incident which happened on this
last evening I ought to mention. We have a
boy, Jack—a likeable lad, as every one says, for
a ten-year-old. I can only say that he is his
father done small. Well, Jack came rushing
into the dining-room to show us an acquisition.
He held at arm's-length the stuffed skin of a
Barn-owl, to which was hanging a ferret. I
looked across to my husband, and said it was
a case of "inherited instinct." He called me a
"Darwinian fool." That night I locked up
the "Glenlivet," and secreted the keys. It was
time.

I want to say that the Grouse cost us thirty-
eight shillings apiece; the Roe-Deer £105.
We were in the Highlands seventeen days.
Last year my bill for dress-making and milli-
nery came to £70. I mentioned these little
facts to my husband. He told me that "the
birds wouldn't lie," "the dogs were not pro-
perly trained" (he gave forty guineas for a brace
of pointers just before we started); that "the

gillies were ignorant of their business," and generally evaded my little effort in arithmetic. Nor is this all. Five days after our arrival at home, my husband "took a Manor." I know what that means :—That the grain is still standing ; that the birds won't come out to be shot; that the turnips are too thick to wade through ; that the "red-legs" (whatever they may be) have driven all the Partridge from the ground. . . .

To-day they have gone " Covert-shooting." Well, I like the plump breast of a Pheasant, and I hope we may have some. But I know. " The trees were dripping ; " " the leaves are still on ; " " the birds were wild."

There ! I'm a poor deluded thing ; and this when I know that the hand-reared Pheasants get fast among their legs.

CHAPTER XIII.

A COUNTRY NATURALIST.

The remarkable race of artisan naturalists so abundant at the beginning of the present century is now nearly extinct. Tam Edwards of Banff is a type of the class; as also Dick, the baker of Thurso. A number of other names at once suggest themselves; but the two first named, through fortunate accident, have now become familiar to their fellow-countrymen.

The country naturalists of which we speak were essentially outdoor and field workers. They came upon the scene when natural science had made but little progress; and each in his hobby, whether geologist or only simple observer of bird-life, did much to place practical physical science on the broad basis upon which it now rests.

Let us take a type of the class. If one were
a botanist he found that the majority of his
predecessors had mostly comprehended the subject as it taught of the herbs and simples of the
wood—

"Ruc, cinque-foil, gill, vervain and agrimony;
 Blue vetch and trillium, hawk-weed, sassafras;
 Milk-weeds and murky-brakes, quaint pipes and sundew."

From this we see how fondly early naturalists
clung to old English names; how they loved
to wrap about flowers the attributes their
fathers had done; and how profound was
their wisdom in "herbalism." There is an
intenseness and simplicity anent their dealings
with nature that is quite captivating. They
were not always infallible observers, however,
and often tripped in their facts. But then,
if natural science has any virtue outside itself,
it is that of which they sought to lay the founda-
tion—the habit of careful and accurate observation. These untutored students of botany
found their science something more, if that
were possible, than a mighty maze; and as to
systematic arrangement there was none.

To the common run of men these bygone
workers were "hobby-minders," and as such

received but scant justice. They were illustra-
tive of types of mental weakness out of which
no good thing could be expected to come. To
show the difficulties with which they had to con-
tend, we will take the case of a characteristic
individual who eventually surmounted every
obstacle. Not only were his pet projects mer-
cilessly reviewed and summarily condemned
because he was " mentally weak," but he sinned
in a far worse way. He was a baker ; and,
after burning the loaves and then underdoing
them, he threw up dough in disgust. It was
evident that baking and botany could not go
hand in hand. This period of his life was the
darkest of all. He was often frustrated and
always condemned.

Our botanist was born in 1702, in one of the
most primitive dales in the country, and which
must have proved a very paradise. Even now
it is one of the most secluded spots in the North.
Two subsequent scientists of renown knew of
its naturalistic treasures, and frequently visited
it. But there was something in the isolation
of this remote vale that told severely against
the earlier studies of the author of the first
great work on English botany. In the fulness

of his boyish enthusiasm he roamed over the hills like a partridge. Unfortunately the crass ignorance of the dalesfolk rendered him a prey to the grossest superstition. Wilson * made long journeys, often at night, among the hills and woods, and by the sea. The country folk said that the lonely calling which took him so far afield *might* be honest; but they shook their heads, and some even ventured to say that he was a " wise man "—a dealer in mysteries, and given to dark sayings.

The evil repute which ignorance and superstition caused to gather round Wilson's early life was probably the cause of his removing to the little grey market-town where he ultimately died. To the superstitious notions which prevailed against him and his consequent removal we probably owe his *Synopsis of British Plants.* He had studied long and hard in his native dale, and had found abundant and rare material. But the want of some good guide—some standard work wherewith to verify his specimens—he sorely felt. Instead of this he only possessed an old " Herbal," as full of

* John Wilson, author of *A Synopsis of British Plants* after Mr. Ray's method. 1744.

inaccuracies as superstition; the only systematic arrangement of which was the *alphabetical* order of the plants. It was fortunate,
however, for all succeeding botanists that he
had made considerable progress in his studies
before coming to the provincial town before
mentioned; for we see how deeply this knowledge impressed him; how accurately he describes
the habitats of rare plants; and with what fondness he dwells upon the floral treasures of his
native vale.

It is a somewhat striking fact that many men
who have made themselves famous in the pursuit of natural science have been shoemakers.
Edwards is the latest example, and it is certain
that Wilson was one. An occupation which
affords ample leisure to the mind is one to
which the subject of this sketch would naturally
lean. Wilson's chief gain, however, in coming
to " Greytown " is contained in the fact that it
brought him within measurable distance of good
books; and of the contents of these, with his
ready knowledge and native understanding, he
quickly made himself master. For a provincial
town, that to which Wilson removed stood
out prominently. It had within it several

remarkable workers — workers who towered
above the common run of men. With some
of these it is known he came in contact. The
libraries of the town, too, were good; and
these, doubtless, he found his greatest boon.
The botanical books which he was enabled to
peruse were the standard works of the period,
but none of them so good as the one he was
destined to write. Of the books from which
he derived his theoretical knowledge, and which
most impressed his after-work, we have before
us a list. Now, curiously enough, not one of
these, even by the widest stretch of imagination, can be called scientific. Order is the
first law of science. As yet, as revealed by
these books, there was only chaos. So few of
the ascertained elements of natural philosophy,
as applied to botany, had been as yet collected
as to permit any arrangement of species, in
any permanently (even over a limited period)
nameable order. Our botanist was the mind
born to perceive and exhibit such order. Then
the simplest and most descriptive nomenclature
prevailed, and it was the best.

Wilson pursued his trade as a shoemaker as
his studies progressed. He married, and some

time subsequently relinquished shoemaking
for what has been called the " more lucrative
employment of a baker." The duties of his
trade, however, soon devolved upon his wife.
The instinct was too strong within him to be
subdued by " loaves ; " and it was at this time
that he did his most solid work in botany.
There is abundant testimony that he had
now become so well known as to enjoy the
friendship of some eminent scientific contemporaries.

To come to Wilson's life-work. His book was
certainly the first English one on systematic
botany. It became then what Hooker's
Student's Flora of the British Isles is to-day.
He called it *A Synopsis of British Plants*, in
Mr. Ray's method ; and of the plants treated he
described the character, description, place of
growth, time of flowering, and "physical virtues."
Enlightened as he was, how could he dare at
that time to publish a book on botany without
giving the " physical virtues ? " These latter
characteristics were said to be set down " according to the most accurate observation " and
" the best modern authors ? " To the whole
was added *a botanical dictionary*, a most valu-

able compilation, accurate for the time, and wholly the author's work. The Synopsis was published in 1744 "by John Gooding, on the Side," Newcastle-on-Tyne. Although Wilson's book was in "Mr. Ray's Method," the author might, in common fairness, have suppressed Ray's name; and it may be remarked that in Wilson's work there is exceedingly little of the "best modern authors" and very much of himself.

This publication was a strong and original work—a very monument of acute observation and the genius of hard work. Viewed in the light of modern science, it contains many blunders; it was published a hundred and fifty years ago, and in this respect is past criticism. The author set out with a well-defined plan, and executed it admirably. His first great object was "to instruct beginners in the knowledge of British plants." Other than works on "Herbalism," which only constituted valuable volumes to burn, botanical works for the most part were written in Latin. Consequently these were closed books to those who were only able to read in the "mother tongue." Essentially one of the people, and having him-

self suffered in this way, Wilson determined
that his researches should be open to all. It were
not otherwise than that he should look with
contempt on most of the work of his contemporaries. There was no regular method, some
disposing the plants in alphabetical order;
others according to their supposed virtues;
though none by any chance giving the student
a ray of enlightenment as to how to identify
any rare plant he might find. The converse
of this Wilson attempted to do, and succeeded.

In classifying plants into their natural orders
and genera, he seized upon one or two well-
marked and permanent characteristics; nor
could he be turned aside by minor differences
or special modifications. And this was the key
to his success. He reduced order out of chaos,
and mightily simplified the work of succeeding
botanists. A feature of his book is that of
giving the habitat of each plant described, and
he seems to derive peculiar pleasure in this.
We have already testified to his accuracy of
description; he is always clear though sometimes quaint. The "herbalist" portion of the
book, which is almost always ridiculous, would
seem to have been dragged out of him and

inserted against his inclination. Of Wilson's method of description here is a specimen— crude enough, it must be admitted, but still sufficiently clear to enable the student to recognize the actual specimen :—

" Wake-robin, cuckoo-pint. By hedges and in shady places. On the top of the stalk grows a long membranous sheath, of a greenish colour on the outside, and purplish within; in which is enclosed a long naked purplish cylindrical pointal, whose lower part is encompassed with a circle of shives. The pointal and sheath fall off, and are succeeded by a thick cluster of yellowish red berries."

And so Wilson wrote the first great work on English botany. To-day his name is hardly known in his native town. He died in 1751. A short time previous to this he had made a strange remark, embodying a stranger request. To a gravedigger he was wont to pass in his daily walks, he observed : " If I have done little good during life, I desire to be of use after death; let my body fill up this hole." The petition was observed, Wilson being buried where he had indicated.

CHAPTER XIV.

NUTS.

Now that nature's duns and browns and yellows have picked out the hedgerows, the nuts have become embrowned at their tips. In the woods the shaggy clusters hang in picturesque profusion, and here and there cause the branches to droop by their weight.

A hazel copse in September is the very embodiment of autumn. The field of golden grain which lay by its margin like a great patch of sunlight on the landscape has given way to the stubble, and the brown Partridges feeding upon this give another characteristic touch to the season. Where the nut-bushes stand out to the sun the nuts are already rosy; deeper in the woods they are still of the green of the leaves. It is pleasant to watch the

progress of the nuts; and as we stand among
the bushes our memory carries us back to the
early days of March. Then the pendulous
catkins swung in the breeze, and wafted their
golden pollen broadcast. Nature is prolific in
her waste, and here was a capital instance.
The almost microscopic red female flowers
started from beneath the scales of the buds,
and were admirably adapted for retaining the
fertilizing pollen. Their slender pink filaments
were covered with a viscid substance, and whenever the bright yellow dust shook against them
it was held fast.

But as March blustered through her moon,
the sun gained in power, the snow began to
melt from off the fields, and patches of green
came through. Then were only white lines
along the fences. The Throstles burst into song ;
the anemone lifted her frail form ; even the pale
primrose peeped from beneath last year's dead
leaves. And one morning as I walked through
the woods a trio of willow wrens told me very
plaintively, and very prettily, that if I pleased,
spring had really come. As though to confirm
what the birds told me, the field-workers began
to turn up the warm, brown land, and a few

nights after the green and yellow catkins all dropped from the hazels. They fell upon the dead leaves, and, raking away these, buried everywhere beneath them were last year's brown nuts. In the larger end of these was a hole, regular as though drilled with a wimble. The sweet kernel had been extracted by one of my woodsy friends—which I am not quite sure, though I will detect him next autumn.

The companions of my woodland haunts are the Nuthatch, Dormouse, Wood-mouse, and Squirrel. A sackful of empty shells might have been counted within a few yards of where I stood, so thickly were they strewn about. But this bright spring day had nothing to do with the empty past, and everything seemed opening to the warmth of the returning sun.

A month later, when the catkins had decayed, the tree had another glory—it was clothed in bright green leaves. The buds might have been seen unfolding to the light, until their tender colour kissed all the woods. The floor of the copse had only a thin layer of soil, but the sun warmed it, and a carpet of blue bells reflected the azure of the sky. A breadth

of green is always delicious, however green;
so is a belt of blue. These are the colours of earth
and sky, and are never monotonous to the eye.

The Rockrose sprang up, and the Wild Thyme
and the Primrose carpeted the woods. Still the
sap ascended, the hazel leaves drank in the
moisture, and expanded more and more. By
June their green rounded lobes were perfected,
and then the hazel was not an unhandsome tree.
It was now, during the time of the running
sap, that the old Basket-maker came to the
copse for the long lissom wands; they are
tough and flexible, and are used for a variety
of purposes. Here, too, the professional Fern-
seller obtained the sticks wherewith to make
his rustic baskets. By this time the soft-billed
summer birds had arrived, and the woods and
copses were flooded with their outpourings.
By the time the old Keeper brought his
pheasants to rear among the bushes the nuts
were beginning to form.

The Squirrel had built its airy nest far out
among the swaying branches of a pine; the
call of a Nuthatch came up from the glade; and
the red Wood-mouse again rustled among the
leaves. Still the nuts grew and shaped until

the clusters stood out prominently among the leaves. The Wood-doves cooed, and had their wicker-like nests among the branches. Moles threw up their runs across the green rides, and at the shadow of a hawk the pheasants ran to the shelter of the bushes. We met armies of Black Ants marching along the paths; saw the Goat-sucker basking among bits of burning lime-stone, and one day discovered that the green corn was just tinged with gold.

The heyday of leafy June was some weeks past, the young Pheasants had grown too large for the Sparrow-hawks, and now the richness of the Hazel harvest was to fulfil the promise of spring.

A month passed, and we looked again. The branches bent, and the rosy clusters shook in shaggy plenty. The hazel copse seemed to centre about it the whole life of the district. Not only animals and birds, but butterflies, beetles, and moths seemed to favour the hazel. The rippling laughter of sunny school-girls rang through the woods, and young and old seemed to be out " A-Nutting."

But there were three little depredators, the cleverest among them all, with which we played

hide-and-seek among the bushes, and watched
their harvesting. One of these was the saucy
red Squirrel. He glided along the branches
like a sunbeam, and constituted our dark-eyed
miracle of the forest. He would watch our
approach, then glide up the high hazel and
survey us from above. Then he perked his ears
and chattered, and once let down a full-ripe
filbert close at our feet. On examining this
we found how he came at its contents, and
often after watched him in the process. He
would sit upon his haunches, half hidden in
the foliage, holding a cluster of nuts. These
he held in his forepaws, and would presently
abstract one, allowing the rest to drop. After
adroitly securing the nut, he quickly rasped
away the small end, and, having made a hole,
inserted his fore teeth and split the shell. He
ate only the largest and soundest nuts, and
was careful to pare off every particle of the
brown skin of the kernel before beginning to
eat. The Dormouse and Fieldmouse adopt a
somewhat different method of coming at the
contents. They gnaw a hole in the shell,
though so small that the wonder is how the
kernel is ever extracted through it.

The pretty little Nuthatch acts in different fashion to all these. With its bright slaty back, orange breast, and black bill, it is quite a handsome bird, and the country children call it "nutjobber." Essentially a bird of the hazel-copse, he may almost always be found there if you invade his haunts. That gnarled and twisted gate-stoop is weathered out of all original conception. There are gaping crevices on its surface, and the Nuthatch has long utilized these. Let us lie here under the bushes and wait for his coming.

Everywhere the half-grown pheasants are feeding—wandering in search of beech-mast, or digging up ant-hills to come at the hidden larvæ. The Dormouse nimbly runs along a branch, and is soon lost in the foliage. There a tiny mouse-like Creeper clings to a lichened bole, and Pigeons are coming and going among the beeches. A Jay chatters from the glade, and flies from clump to clump with its heavy flight and harsh cries. Partridges are scratching among the brown loam; and now and then rabbits pop in and out of the green brackens. Presently the Nuthatch flits to the old gate-post, and fixes a nut in the crevice.

It taps lightly at first, then as though it had fasted for days. Retiring an inch or so, it brings down its bill with the whole force of its body, and bits of the stubborn shell begin to yield. Soon a hole is made, and as pieces of the kernel fly off they are snatched up ere they reach the ground. Again and again it darts, always securing the chips, until the dwindling shell yields up the kernel. Of this the bird is not long in disposing. As soon as it has finished, it flies off to the copse. For yards around the gate are pieces of shell and outer coats of beech-mast, as well as the harder cases of seeds and wild fruits. Upon all these the Nuthatch feeds. The little bird is as interesting a creature as any of the denizens of the hazel-copse.

CHAPTER XV.

CONCERNING COCK-FIGHTING.

THERE is no place like a sleepy country-town for retaining primitive manners and customs. There is a quaint conservatism in everything and everybody about the place. And just as old ale and older traditions are treasured, so are relics of barbarism.

The men at the sign of the "Bush" are addicted to Badger-baiting whenever a "brock" can be found. They keep a "scratch" pack of tiny draghounds to hunt Martens among the crags, and are always hard in the wake of the Otter Hounds in summer. Bull-baiting has gone out, but the ring remains on the green to which the beasts were fastened. This was indulged even at the beginning of the present century, and in many places was upheld by Corporate bodies.

But of all these barbarous sports Cock-fighting was perhaps the most popular. Nor has its popularity declined to any great extent. About Easter-time, assemblages of surreptitious cock-fighters are still pounced on by the police; and a certain class of the community sticks to this cruel sport in spite of all the efforts of the law to suppress it.

Cock-fighting in this country is mostly confined to the North—to the quiet spots among the hills where the chances of molestation are small. The pleasure to be derived from a Main of Cocks appeals to a wider class of the community than is generally supposed. Among the upholders of the sport are farmers, magistrates, and yeomen. But these can point to famous modern patrons who were their immediate predecessors.

When he lived at Elleray, "Christopher North" was wont to carry a game-cock under his arm in his walks abroad, ready to pit against that of any of his neighbours. It is even reported upon trustworthy evidence that the Professor indulged the pastime in his drawing-room on Sunday afternoon. Of course he himself does not tell us these things, but those who have read his life by his daughter, will remember frequent entries in his diary. Thus:

" Red pullet in Josie's barn was set with nine eggs on Thursday; small black muffled hen set herself with eight eggs on Monday." And side by side with some beautiful lines from the *Isle of Palms* is ranged " a lists of cocks for a main with W. and T. ; " then follows an enumeration of the birds. Wilson thought much of his game-fowl, and bred only from the best fighting strains in the country. He probably acquired the " art " at Oxford, where in his time mains were regularly fought.

In the present century there were royal Cockpits at St. James's Park and Westminster, the former being pulled down in 1824. Hogarth's and Cruikshank's pictures show the interiors of these pits, and the motley company which attended them. The mains were fought on raised circular platforms, and Hogarth depicts a man in a cage suspended from the ceiling, suffering the penalty of being a "blackleg." To these pits came peers and pickpockets, butchers and jockeys, ratcatchers and gentlemen ; and gamblers of every description. Not only our own but European monarchs visited them upon occasion, and graced the cocking with their presence.

One of the phases of this ignoble sport was that connected with the Free Grammar Schools of the North. These were endowed with a stipend for the maintenance of a master, but in many cases this was quite insufficient. As the nature of the schools entitled the preceptor to nothing more than his salary, the parents of his pupils thought proper to reward his diligence by an annual gratuity at Shrovetide, called a " cockpenny."

In connection with this institution it is recorded that a singular donation was made to a northern grammar school by a Mr. Graham, a cavalier, of a silver bell, upon which was engraved " Wray Chapple, 1655, to be fought for annually on Shrove Tuesday by cocks." Some weeks previous to that day the boys fixed upon two of their school-fellows for captains, whose parents were able and willing to bear the expense of the coming contest. The master, upon entering school, was saluted by the boys throwing up their caps, and the exclamation, " Dux! Dux!" On the appointed day, the two captains, with their friends and school-fellows, who were distinguished by blue and red ribbons, marched in procession to the

village green, where each produced three cocks,
and the bell was appended to the hat of the
victor. In this way it was handed down from
one successful captain to another. It was not
until 1836 that these contests were discontinued,
and in their place there is now an annual hunt.
The bell of 1655 continued a parochial institution for 217 years, and had long graced the
hats of " cocking champions," or the white rods
of sham mayors.

Although many of the more famous cockpits now exist, they are grass-grown and
unused. With regard to the actual fighting,
the plucky birds were supposed to set an excellent example to virtuous youth; and to
arouse a noble emulation in fighting the Gallic
or other wide-throated cock that dared to crow
defiance or flap his wings. In this connection
it may be added that in some of the endowed
schools before mentioned as much as half the
master's salary had, by arrangement of the
founders, been made to depend upon the cock-
pennies.

Cock-fighting was also popular in Scotland
at a very recent date. With what zeal it
was practised may be inferred from the fol-

lowing : One Sunday, in St. John's Chapel, Edinburgh, an old gentleman was sitting gravely in his seat, when a lady in the same pew moved up, wishing to speak to him. He kept edging cautiously away from her, till at last, as she came nearer, he hastily muttered, " Sit yont, Miss ——, sit yont! Dinna ye ken ma pouch is fu' o' gemm eggs ? "

A parallel to this enthusiast is to be found in an old lady of Houghton, in Cumberland, who on one occasion admitted she had " gone down on her bended knees," and prayed that a certain cock of her feeding might win at Newcastle.

Mention of Cumberland brings to mind the fact that this is the county most famous for cock-fighting both in past times and present. Mr. Chancellor Ferguson says that the sport had episcopal as well as royal patrons, and that a cockpit even now exists at Rose Castle, .the palace of the Bishop of Carlisle. Whether any bishop was ever actually present at a main is not known, but the pit was the scene of many famous battles between the " black-red " and " grey " cocks of the bishop's neigh-bourhood and an adjoining parish.

In Cumberland, cocks were commonly fought

within the pleasaunce of the palace, sometimes within the churchyard itself. Occasionally this went on while service was proceeding, though some had sufficient regard not to begin until the conclusion of the sermon. At another town in the same county, the grammar school-boys who were the winners of the mains had presented to them Prayer-books, " with inscriptions suitable to the occasion ;" probably some stirring reminiscence of the sport. At the present day the seal of the Dalston School Board displays a " black-red ; " but the motto, " While I live I crow," is omitted.

A famous northern cocker has, during the present year, furnished the following particulars on training cocks, to the President of a Northern Archæological Society :—Now these cocks were taken from their walks, say to-day, Friday, and fought about Monday or Tuesday week. Say the cock was five pounds weight, or a little under, at the time he was taken up, he would fight four pounds four ounces or so. The first part of their training was to cut a little off their wings and tail, give them senna tea to drink until (say) Tuesday, cut their spurs short, and spar them every day with small boxing-gloves tied on their heels. On Tuesday they got their medicine—the very best Turkey rhubarb and magnesia, about the thickness of your first finger, in fact, more than would quickly operate on you or me ; next day senna tea and sparring. They got very much reduced by Friday—all the fat out of them. After that they give them new milk, and bread made of eggs, loaf sugar, etc., in fact, everything that is good, the very best malt barley, and so on. For the old cockpit they used to feed at different public-houses—one was in Pack Horse Lane, another in the Castle Lane, in fact all the lanes in English Street, Carlisle. They fought single battles for £5 or £10, and what they called four mains, that is, four cocks ; of course, the winner had to get two battles.

CHAPTER XVI.

A PINE-WOOD STUDY.

The dying sun sends a blaze of purple light
and throws a lurid crimson over the shaggy
pine trunks. The Cock of the Woods crows
from the pine-plumed gloom, and the light
shoots upward. Then the tall columns range
themselves into aisles, and nought but silence
possesses them. Deep depths of pine-needles
have blotted out all fair vegetation, and the
genii that guard the forest solitudes are great
Eagle-owls, for ever night-haunting in famine
for prey.

Rugged and corrugated bark covers the floor;
and a fallen monster has crashed through the
branches of surrounding trees. For the Pine,
even when set amid its own wild heather, is
never deep-rooted. On the bare brae, its roots

wander in the wildest reticulation over the grey rocks. The lichened trunk towers high up to its umbrageous canopy, and adds its picturesqueness to the wild scene.

The light of day reveals new life, and dispels the gloom. There are open spots in the forest which are as oases in the desert. Sunflashes come·to the glades, and the glossy-plumaged Wood-doves coo in them. In the pine tops they have their platform nests—wicker-like—each containing two white eggs. The Wood-pigeons not only nest here, but have their regular roosts through the greater part of the year. The red-furred Squirrels look for you, hide from you, though always with a branch intervening; they have their " dreys " in the angles of the boughs. Squirrels have two nests. The summer one is a slight structure, swaying far out on the frailest twigs; in this the young are produced, and its aërial site acts as a protection. The winter nest is against the trunk of the tree, and is thickly lined with hay and soft needles.

Through the paths and rides of the forest the pronged Roe deer wanders in spring and summer, feeding upon the succulent vegetation of the sprouting woods. The woods lie on the confines of the forest, and are open and broken. At night the deer visit the water-springs, and in the bracken-beds drop their soft-eyed young. They are within hearing of the belling of the Red deer on the corrie, and some-times wander to the grey lichen patches loved by the Ptarmigan. But the Roe deer is essen-tially of the woods, the Red deer of the mountains. In winter the Red Grouse come to the sheltered forest, and through all seasons the noble Black-cock crows from the sunny brae to his grey-hen in the hollow. All these children of the mist suggest a wonderful fitness to their home. The Red deer and the Roe are indigenous, as is the Red Grouse; and this applies equally to Ptar-migan and Black-game.

The only fit setting for the Pine is that of the sublime and majestic solitudes which have produced it. In this combination there is per-fect harmony; or at least in those Highland districts where the Golden Eagle still yelps as it flies down the corrie, where the Salmon leaps in the burn pools, and the Red deer bells from the hills. For even to-day such spots there are. Of the wild mountain and moorland tracts, the

I. Hutchinson lith.

G. E. Lodge

West, Newman imp

Golden Eagle

pine forest seems to centre about it the life of the district. In it the corbie crow builds, and on its confines may be heard daily the hoarse croak of the Raven. There is no tinge of superstition in the fact that this sable saulie in his overflow of animal spirits may be heard to laugh, as he goes dallying round and round some dead Herdwick. Upon his steel-blue legs he whets his formidable bill, and then, by cut and tierce, digs at lip, palate, brain, or nostril. And the "majestic" Eagle, monarch of the glen and foul feeder at once, hastens to enjoy the meal. This is hard fact, and the bird-king at times is but a great "carrion" crow.

In spring, the rare Dotterel comes to the hills above the pine zone, and breeds far up among the mists. The Buzzard still remains, and a few pairs of Peregrines nest among the rocks of the crags. There is lesser life in the pine forests themselves, both bird and insect; this abounds mostly in winter. The Coletit searches the pine-needles for cocoons of insects, and flocks of Goldcrests and Siskins together range the woods. When the weather is hard, and the pines snow-plumed, a flock of restless chattering Crossbills sometimes makes its appearance to feed upon the cones. The admirable bill of the bird enables it to do this, and split open the hardest seed-cases. Hanging in every conceivable attitude, the birds use their crossed mandibles, keeping up a perfect shower of cones.

As well as to birds, the balm and gum of resinous woods seem to have a charm for a great variety of insects. Most of these visitors are injurious to trees, and feed upon products which would be speedy death to others. The Giant Sirex is a hornet-like borer, which does much harm to growing timber. The grub bores to its retreat, feeds upon the hard wood until its pupa state, then emerges a perfect insect.

The pretty Pine-moth has its wings orange-brown, variegated with rich dark-brown and grey. Its larvæ feed upon the fir, and may be taken in great quantities. On some bright spring day the cocoon opens, and the pine beauty emerges. Then it basks on the bark, opens its wings to the sun, and it is only whilst moving that the insect can be detected. The rare Pine-hawk Moth is a product of the forest, and a beautiful insect. But these, like the brightly coloured fungi of

the woods, are parasites, and riddle and bore
and drill the timber in all directions. Borings
like those of the *Scolytus*, tunnellings under the
bark, and the formation of resinous galls are
all insect doings. Fairy-like fungi in a variety
of forms is parasitic on the pine, and relieves
the somewhat sombre colouring by flashes of
scarlet and glowing lustre. Beetles and insects
innumerable swarm through the fallen needles,
and sport their gauzy wings in the sunlight.

To the open spots, where the Jays and
the Cushats come, a patch of forest flowers
and bright green grass glisten. Even here,
on the thinly soiled ground, the golden
Rock-rose blooms, the Wild Thyme, and the
Tormentilla. Sometimes even the Giant Bell-
flower rears itself, and a few species of Carex.
This bright green caterpillar, with brown-black
stripes, will first change into a brown chrysalis,
then into a beautiful Pine-hawk Moth—not
a common entomological find. In addition to
these, in and about the wood, are the Black
Arches, Barred-red and Grey Carpet-moths, the
Light, Silver-striped, Scarce Orange-spotted,
Spotted and Streaked Pine, Scarce Ermine,
Large and Resin-grey moths, and a host of

others.   Many of these are beautiful, and all interesting.

The Pine it is that gives us the wood called " deal," the most valuable of European timber. It is used for a variety of purposes in both civil and naval architecture.   Our limited forests cannot in anywise meet the demand, and tens of thousands of magnificent trees are transported annually to our shores.   This supply comes from Norway and Northern Russia.   But young forests are rising in our land, and soon will rob it of its barrenness.

CHAPTER XVII.

" OLD KITTIWAKE," as he was invariably called, was a survival of pre-breechloading days. His decline was contemporaneous with the improvements made in firearms and all relating to shooting. In winter he lived his lonely life on the mosses and marshes; but during summer he turned from fowling to fishing, or assisted in the game preserves. His outward garb seemed more a production of nature than of art, and was only changed when, like the outer cuticle of a snake, it sloughed off. When any stray Fowler or Shore-shooter told Kittiwake of the effect of a single shot of their big punt guns, he would cap their stories by going back to the days of decoying. That was the time for fowl! and there was pathos in his voice as

he told how the plough had invaded the sea-
birds' haunt, and explained that what was now
corn-land had once been a Gullery.

In his youth, Kittiwake had "worked" a
duck decoy, and his father and grandfather
had been fowlers before him; even now he
surrounded his craft with as much mystery
as does the fish-poacher his preparation of
salmon-roe. The old man was wont to tell of
clouds of Widgeon, banks of Brent Geese, and
the "demon huntsman" with his Wish Hounds
whose cries came from the leaden wintry skies.

These are Wild Swans coming from the
icy north, which choose dark nights for their
migrations. The grand trumpet notes are
sounded clear, distinct, and clarion-like; as
a solitary bugle sounding the advance; or,
"like the tongue of some old hound, uplifted
when the pack runs mute with a breast-high
scent; then, as if in emulation of their leader's
note, the entire flock bursts into a chorus of
cries, which, floating downwards on the still
frosty air, has every possible resemblance to the
music of a pack of fox-hounds in full cry "—
sounds which have doubtless given rise to old
Kittiwake's legend.

Kittiwake was full of strange lore about
birds, and the wild sea-fowl and waders he
loved best. He remembered the breeding-
haunts of Ruffs and Reeves, and would tell of
their strange fights at pairing. When he was
a lad, Bitterns not only boomed in the bog, but
bred there ; and he had once, though only once,
seen a Great Bustard. It was in a field of
young wheat, and stood as high as one of the
fawns in Honeybee Woods. The 'Squire shot
the great bird, and the folk flocked to see it
from miles round.

In his own rough way he was an admirable
naturalist, and whenever a more than usually
rare wader struck his nets he was careful to
preserve it. He was a complete master of the
art of setting gins and springes, and the primitive contrivances he had for taking shore-birds
were almost as numerous as the species for
whose capture they were designed. Indeed,
his great success lay in the fact that he was
a close and accurate observer. He carefully
noted the haunts and habits of wild-fowl, then
set his nets and hair-nooses by the light of
his acquired knowledge. In winter, Snipe
were always numerous on the mosses which

margined the coast. The Snipe is one of the first birds to be affected by severe weather ; and if on elevated ground when frost sets in, immediately betakes itself to the lowlands. At these times Kittiwake had an ingenious method of taking this pretty little game-bird. This was by means of a "pantle" made of twisted horsehair. In preparing his snares the old fowler trampled a strip of oozy ground, until, in the darkness, it had the appearance of a narrow plash of water. The Snipe were taken as they came to feed on ground presumably containing food of which they were fond.

As well as Woodcock and Snipe, Larks were taken in the pantles by thousands. These were set somewhat differently to those intended for the capture of the minor game-birds. A main line, sometimes as much as a hundred yards in length, was set along the marsh, and to this at short intervals were set a great number of loops of horsehair in which the birds were taken. During the migratory season, or in winter when larks have flocked, sometimes a hundred bunches (of a dozen each) would be taken in a single day.

During the rigour of winter, too, great flocks

of migratory ducks and geese came to the bay, and prominent among the sea-ducks were immense flocks of Scoters. Often from behind an ooze bank did we watch parties of these playing and chasing each other over the crests of the waves ; they seemed indifferent to the roughest sea. The coming of the Scoter brought flush times to the fowler. Kittiwake would not unfrequently take half a cart-load of these black ducks in his nets in a single morning.

The birds feed upon mussels and soft bi-valves, following the advancing tides shore-ward in search of them. These facts Kittiwake noted, and worked accordingly. He carefully marked where the birds fed, saw their borings and stray feathers, and then, when the tide ebbed, spread his nets. These were fixed by pegs at each corner, and raised about fifteen inches above the sandbank. Returning to feed with the tide, the ducks dived head-fore-most into them, and became hopelessly fast. Another of the sea-ducks, the Scaup, he took in large numbers in the same way. Some-times a lovely Velvet-Scoter, or rare Surf-Scoter, would be among the spoil, but these occurred at long intervals. So great were the

numbers of Black-Headed Gulls which once
nested on the mosses, that Kittiwake used
to feed his old shaggy horse upon their eggs
for two or three weeks during the breeding
season. Morning and evening he collected a
basketful, and so long as the eggs lasted,
Dobbin's coat was always soft and sleek. Years
subsequent to this the plough invaded the sea-
birds' haunt, and now a small town occupies
the site of the once famous Gullery.

In addition to his nets and snares, the only
fowling-piece that Kittiwake ever deigned to
use was an old flintlock with tremendously long
barrels. Sometimes it went off; oftener it did
not. I remember with what desperation I
upon one occasion clung to this murderous
weapon whilst it meditated, so to speak. It is
true that it brought down quite a wisp of
Dunlins, but then there was almost a cloud of
them to fire at. These and Golden Plover were
the game for the flintlock, and the old man was
peculiarly successful. He was abroad on the
marshes at dawn; and at that time plover fly
in close bodies or feed in the same relation.
Sometimes a dozen birds would be bagged at
one shot. But the chief product of Kittiwake's

night work on the marshes were obtained by
his cymbal-nets. Wild birds were brought
down by decoys, and when the former came
within the province of the net it was rapidly
pulled over and the game secured. For the most
part only small birds were taken in this way.

Numbers of Mallard and Teal bred on the
marshes in summer, and there was also a
colony of beautifully plumaged Shelldrakes.
In August and September the old man
captured immense quantities of "flappers"—
plump wild ducks, too young as yet to take
wing. These were either caught in the pools
or chased into nets set to intercept them. From
these Kittiwake derived an important part of
his revenue, as they invariably brought good
prices in the market.

It may not be a matter of common knowledge
that there is a short period in each year in
which fully matured wild ducks are unable to
fly. The Mallard especially is an early breeder,
and soon after the brown duck begins to sit the
male moults the whole of its flight feathers.
So sudden and simultaneous is this process that
for six weeks in summer the usually handsome
drake is quite incapable of flight; and it is

probable that at this period of its ground existence the assumption of the duck's plumage is
a great aid to protection.

The Shelldrakes were the handsomest of the
wild-fowl on the marsh. A colony occupied
a number of disused rabbit-burrows on a raised
plateau overlooking the bay. The ducks were
bright chestnut, white, and black, and laid from
eight to a dozen creamy eggs. As these really
handsome ducks brought large prices for stock-
ing ornamental waters, Kittiwake used to
collect the eggs and hatch them out under
hens in his turf cottage. This was quite a successful experiment up to a certain point; but
the ducks, immediately they were hatched,
seemed to be able to smell the salt water, and
would cover miles to gain it. With all the
old man's watchfulness, the downy ducklings
sometimes succeeded in reaching their loved
briny element; and, once in the sea, they were
never seen again.

Among the strange birds which the old
man had known as resident on the Marsh were
various of the rarer Sea-swallows, and especially
he was wont to talk of the curious Ruffs and
Reeves. He had many marvellous stories of

how they fought in spring time, and when they were plentiful he used to take them in nets, and fatten them on soaked wheat for market. Even from this bleak northern marsh the birds were sent by coach all the way to London. By being kept closely confined and frequently fed, in a fortnight they became so plump as to resemble, when plucked, balls of fat; and they then brought as much as a florin apiece. If care were not taken to kill the bird just when it attained to its greatest degree of fatness, it fell rapidly in condition, and was nearly worthless. The marshmen were wont to pinch off the head, and when all the blood had exuded, the flesh remained white and delicate.

Held in even greater estimation as delicacies than Ruffs and Reeves were Godwits, which were fattened in like manner for the table. They were more rare than the last, and fetched higher prices. When a bunch of birds was sent to town, a basket of pink-fleshed Char was invariably sent with them ; and, owing to the rarity of this fish, Kittiwake was always able to sell them well.

Coots came round in their season, and

although they yielded a good harvest, netting them was not very profitable, as only the villagers and fisherfolk would buy them. Their flesh was dark and " fishy."

Experiments in fattening were upon one occasion successfully tried with a brood of Grey-lag Geese, which the old fowler discovered on the marshes. As this is the species from which our domestic stock is descended, he found little difficulty in herding, though he was always careful to house them at night, and wisely pinioned them as the time of the autumnal migration came round. He well knew that the skeins of wild geese which at this time nightly crossed the sky, calling as they flew, would soon have robbed him of his flock. To give some idea of the enormous number of wild-fowl which once visited the fens and marshes, it may be mentioned that a flock of wild ducks has been observed passing along from north and north-east in a continuous stream for eight hours together. But this was in the good old times. Kittiwake and his occupation are alike gone.

CHAPTER XVIII.

RARELY has the grain been so golden or the wild fruit so plentiful as now.* The berry-bearing bushes are wreathed with scarlet loads, and bend beneath the heavy fruit. First and fairest among them is the Mountain Ash or Rowan tree. Everywhere it hangs out its clusters of orange fruit to the sun. It conforms with an easy grace to the heather brae, and in its wild situation is loved by every one. All the birds flock to it, and the Ring-ouzels never leave it so long as there is a berry remaining. Taking advantage of this fondness for its fruit, bird-catchers bait their hair nooses with the berries, and hence one sometimes hears the Rowan called the fowler's service-tree.

* Autumn, 1889.

Nowhere are old-fashioned hedgerows so common as in England, and the most prominent fruits of these are hips and haws. "Many haws, many snows," is a proverb which associates a hard winter with an abundant crop of this particular fruit. The Hawthorn is always beautiful—first with its snowy or pink blossoms in spring, and now with its variable fruit. Sometimes this is woolly, then golden yellow, yet again of a bright deep red—a gaudy advertisement which the tree hangs out to many a hard-pressed bird.

Only those learned in botanical lore know how many species of wild roses there really are in England. Some are better known than others, and one way of distinguishing them is by their fruits. The White Dog-rose has bright red berries, globular in shape, and mounted on purple foot-stalks. Then there is the Common Dog-rose, with its flask-shaped scarlet fruit; and the hips of the Sweet-briar, which are elliptical.

Nearly akin to this—the Eglantine of the older poets—is the Woolly-leaved rose, a beautiful species, the whole plant being covered with glandular hairs, which emit a delicious aroma. This odour and the hairs are also characteristic of the deep purplish-red fruit. " Conserve of hips" will be known to all as the prepared pulp of the wild rose. The Sloe, or Blackthorn, also belongs to the family Rosaceæ, and is just now hanging out its sprays of globular black fruit covered with delicate bloom. Before this is sullied it is of the most inviting appearance, but its good looks are sadly belied to the taste before the mellowing influence of the first frosts. The Sloe tree connives at adulteration, for not only does its " wine " impart a deeper crimson to our cheaper " ports," but its dried leaves are used in compounding tea; and hence it has been said that " the simple sloe of our hedges is verily a plant of wonderful endowments, since Chinese Bohea and the wines of Portugal are equally procurable from its beneficent branches."

The Wild Service-tree is beautifully flowered, beautifully fruited, and the clusters of white May blossoms are replaced in autumn by pro-fuse bunches of brown berries, which, if acid at first, lose their rough flavour by frost. Then there is the Whitebeam of the limestone escarp-ments, with its sharp contrasts of white and scarlet, which hangs out such a feast for the birds. It is said that the Hedgehog is extremely

partial to the mealy pulp of Whitebeam berries,
and will make long journeys in search of them.
The Medlar is more a "fruit de fantaisie"
than of utility or beauty, though this cannot
be said of the Guelder-rose or the Wayfaring-
tree. The former has clusters of brilliant
red berries, the hidden parts tinted with rich
yellow, and having an exquisite semi-trans-
parent waxen texture. The Snowball tree of
our gardens is also a variety of guelder-rose.

In spite of the Yew being a "cheerless un-
social plant that loves to dwell among skulls
and coffins, epitaphs and tombs," it puts forth
a brave show in autumn, and its scarlet
berries stand out sharply against its dark-green
foliage. The poisonous properties of the Yew
are well known, though this does not apply to
the berries. These are sweet and viscid, and
are readily eaten by birds and children. The
ripened scarlet fruit of the Barberry may often
be seen covered with masses of orange-coloured
dust like pollen, resembling "rust" in wheat.
This is a tiny fungus which shows up beautifully under a low power of the microscope.
The berries of the Woodbine are just now
showing their various tints of green, orange,

and yellow, but will soon assume that bright cornelian red which renders them so ornamental in the hedges.

Now that the hedge-rows are bare, the tendrils of the Honeysuckle show that curious spiral growth which is so constant in one direction— from left to right; whilst the neighbouring Black Bryony twines in a precisely opposite direction. Blackberries hang in luscious bunches by the roadside, and the Elder-trees swarm with birds in search of their purple-black berries. It is almost impossible to estimate the economy of the Elder in the country. Every part of it is used—its cream-coloured flowers, its berries, its bark, its leaves, its wood; and the old pharmacopœia is full of allusions to it.

The myrtle-like Privet has the power of conforming to almost any surrounding, and in widely different situations is now putting forth its rich masses of purple-black berries which stand out so vividly against the green foliage. Then there are varieties of the fruit—white, yellow, and green—which contrast well with the normal colouring. All the winter birds love the Privet, feasting upon its berries far into the fall. Just prior to dissolution, the leaves of the Dogwood or wild Cornel tree have passed through their various shades to a deep red colour. These vie with the dark purple berries, which though bitter and astringent to the taste, are pretty to look upon as they hang in shaggy clusters.

The Spindle-tree in autumn is one of the most beautiful, and although its fruit is not a berry, the curious pendent seed-vessels which hang upon the tree give it a very delicate appearance. These rosy capsules have a soft waxen texture, and in the autumn winds quiver prettily among their long footstalks — a fairy creation in the flowerless season. Then in the hedge-rows is the Holly, with its glossy leaves and scarlet berries, and many other fruit-bearing shrubs with more or less conspicuous seeds. Among these may be mentioned the Buckthorn, Bird - cherry, and the curious Butcher's Broom.

If the woods and hedges produce all these, what may be said of the heaths and moorlands? Among the low-growing shrubs we find upon the commons and mountains a host of beautiful berry-bearers and their near neighbours. Among these are the Bilberry and Blaeberry,

purple Heather and Ling, the Mount Ida or
Whortleberry; Cranberry, Bearberry, Crowberry, and Juniper; and the Sweet-gale or
Candleberry. Then there are the Cloudberry,
Dewberry, the low-growing brambles, and a
host of others. Of these one of the best
known is the Bilberry, with its rosy waxen
flowers, fresh green foliage, and its prettily
bloomed globes in autumn. The wiry shrub
which bears this fruit ascends to over three
thousand feet, and lower on the slopes various
of the game-birds are fond of it. A fruit like
the last is the Bog-whortleberry, which also has
a glaucous bloom, though it is larger than the
last.

Perhaps the prettiest of this group is the
Mount Ida or Whortleberry—pale pink and
coral at first, then scarlet as the season advances.
But the best known of all is the Cranberry, so
much of the rose-coloured fruit of which is
brought from northern bogs and heaths to
southern markets. The Crowberry has shining
black fruit, and, like so many of the species
named, is sought after by moorfowl. The Cloudberry is one of the low-growing brambles, as is
the Dewberry. The gray-green Juniper bushes

of the limestone districts bear large quantities
of close-set berries of various colours, and these,
unfortunately, constitute a commercial com-
modity of a very questionable kind, and are
much used in connection with the manufacture
of gin.   This by no means exhausts the list of
autumn berries, but the remaining ones are
neither so conspicuous nor so interesting as
those enumerated.

CHAPTER XIX.

A GREAT BIRD-FIGHT.

AT Hernwood exists one of the largest Rookeries in the country. Hundreds of Rooks annually nest there ; and vast flocks, counting thousands, fly to and fro morning and evening. At times other than the breeding season the tall trees of the wood constitute one mighty roost.

At the grey of morning the birds go forth with loud clamour and caw, and are dispersed over their feeding-grounds before the mists have rolled from the valleys. They are abroad thus early in search of wireworms and larvæ. The freshly ploughed fields in spring attract them even before light has come. Their return to roost is more regular than their dispersal. At evening—in nesting-time, often after dark —the rooks in long lines fly down the valleys,

cawing as they go. Each clanging file is led
by some "many-wintered crow," which never
deviates from a well-defined route. When the
valleys converge and the flocks unite, aërial
evolutions are indulged in which show the
wonderful wing-power of the bird. Especially
is this so flying "down wind." When all
are assembled, the snapping of sticks and loud
cawing becomes general. These gradually
subside; although far into the night an occasional caw comes from the tree tops or a falling
twig startles the stillness.

At Hernwood, too, is a Heronry—an historic
one—the once royal game-birds having had
possession time out of mind. But since the
degeneration of the majestic fisher the number
of Herons has decreased. They have not that
protection accorded them now that they had as
royal quarry—quarry worthy to be flown at by
princes. But still, at Hernwood, they find an
asylum; and the thirty remaining nests produce
annually about one hundred and twenty birds.
These, from February till October, haunt the
Heronry, betaking themselves for the winter to
the mosses and marshes; and some few—like
Wordsworth's immortal leech-gatherer—wander

from pond to pond, from moor to moor. Individuals poach the trout-streams; while the majority—gaunt, consumptive, and sentinel-like —stand along the channels waiting for the flow.

When the sun shines the Herons droop their wings and the sand-banks are lit with a bluish-grey haze. As the wind gets up, the birds repose their long necks, depress their crests, and stand upon one leg. But the tides surely flow, and as surely resolve the Stacy-Marks-like group into animation. The Herons fly low over miles of channel. There are flooks and flat-fish to be fought over with the Lesser Black-backed gulls; there is stealthy wading to be done; and woe to the fish that comes within range of that formidable pike! No aim so unerring as that of Heron, no poacher so successful! And thus, with crest erect and every sense acute, does our angler pursue his solitary trade.

A hundred Herons live in Hernwood with thousands of their sable neighbours—in amity for the most part, with only occasional feuds. But this was not always so; for just over a century ago—in 1775—a memorable fight took place which lasted three days. Hundreds of Rooks were killed, as were scores of Herons.

Individual skirmishing had occurred annually, but never had possession of the groves become a party question until now. A crisis had come, and every bird stood to its colour—slaty-blue or black. And this was how it came about.

In a grove of fine old oaks the Herons had lived and bred—their right, possession time out of mind. The oaks were felled in the spring of 1775. Great was their fall; for they were hoary and heavy, and at this time they contained the nests and eggs of scores of Herons.

The birds were disconsolate for a time, but soon sought to found a new settlement. The time of second nests was at hand. Near their old habitation only young firs grew, and these were not substantial enough to contain their bulky nests. At this the Herons, determined to effect a standing, invaded the haunts of their neighbours. They met with an organized and stubborn resistance; but, although their sable neighbours greatly outnumbered them, they were so far successful as to found their colony.

It is with the fight in which the herons came off successful, however, that the interest of the Hernwood episode attaches. This, as before remarked, lasted three whole days, and upon each was carried far into the night. Naturalists traversed long distances to witness the novel spectacle, and one eminent among these averred that a battle on such a scale was unique in the history of bird-life. Although the number of dead Herons was as nothing to the number of Rooks, yet the former suffered severely. Vastly outnumbered, it was only by accident that a Heron was stunned by the strong bills of its opponent, and was then done to death. The Rook that came within range of heron's pike had its skull pierced, and death was instantaneous. And so, at the beginning of the fourth day, the fight began to lull from sheer exhaustion of the combatants. Dead and dying birds continued to fall from the trees for days; but the fight was at an end. Not only were the Rooks driven away, but the Herons captured the trees and successfully built their nests. During the after-building there were fitful outbursts, but these came to nothing. Incubation proceeded in due course, and that year two broods were reared by each pair and successfully carried off.

In 1776, the fight was renewed on a much smaller scale; but the Herons, retaining their trees, again came off victorious. Of the superior

fighting power of the latter the Rooks seem to
have been at last convinced; they abandoned
the grove seized upon by their neighbours.
And now the Herons, naturally peaceable, live
in perfect harmony with their more noisy
neighbours. Such is the historic feud of the
bird clans at Hernwood.

To-day the Herons occupy the highest trees
in the most elevated part of the wood. These
are of three species—ash, elm, and beech. The
nests are large cumbersome structures, built of
boughs and lined with larch twigs; the pale-
blue eggs rest in a slight depression. When
the trees are swayed by the wind the eggs
sometimes roll out; and we have frequently
seen young herons which have been blown
down, stalking among the trees. Now it is not
at all unusual to see nests of Rooks and Herons
in the same tree. The birds never steal each
other's sticks.

As the Heron has two, occasionally three
nests yearly, its breeding-season lasts from
early spring to late autumn. Then the young
are taken to the marshes and along the channels. They feed upon almost every species of
fish and numerous crustaceans. As surely as a

young heron is seized, so surely does it disgorge—an eel, a toad, or even a water-vole.

Herons are as omnivorous as mankind itself, and take various small animals when opportunity offers. They are also endowed with wonderful powers of assimilation and digestion. On the whole, the Heron is among the most interesting of birds; and concerning it there is yet much scope for original investigation. It is a poacher of no mean merit; and we have seen spawning-streams covered in to limit its depredations. Its destruction of coarse fish and noxious water-larvæ is great in comparison with any trout-fry it may destroy. Its good qualities far outweigh its bad ones, and it is a bird to be protected and encouraged.

CHAPTER XX.

WINTER BIRDS.

JUST when the great army of Warblers that have charmed us through summer-time are preparing to leave our shores, a like bird-movement is setting in towards us. Our departing guests are comprised mainly of the soft-billed wood-birds, but the host of winter visitants is made up of hardier forms.

The migrations of these feathered things are still among the mysteries of science, but every year we learn some little about them. We know now more accurately the lines and times of migration, and as to the manner in which the flight is conducted. Recently the British Association has made some accomplished student of bird-migration responsible for accurate observation over a well-defined area, he in turn being

assisted by an army of willing helpers, composed
of the keepers of Lighthouses round our coasts.
It is noticed that thick and foggy weather
marks the period of the heaviest migrations;
and that the great autumn movement is performed in one or more vast " rushes." Strange
it is that the aërial journeys of the little
migrants are invariably conducted in the darkness and against a head-wind. As though
following some long-lost land-line, the birds
regularly take the sea over well-defined tracks.
The autumn immigrants fly from east to west
and north-west—their return journey being
conducted over the same lines, though in a
contrary direction. It has been noticed that
the first great flight occurs about the middle of
October, the second just as regularly a month
later.

Perhaps the best way of observing the spring
and autumnal bird-movements is to set one's
self right in the track of the migrants. It is
marvellous how such frail things as Goldcrests
make head against a storm, yet the following shows how vast migratory flocks of these
tiny creatures encounter their perils. Autumn
winds have torn the more brittle boughs from

every tree. Shaggy and lichened bark covers the pine wood floors. Upon depths of pine-needles other needles drop, blotting out all fair vegetation. Yet the deadness and dreariness of these tree-tracts have become animate for a while, and from every bough and crevice come the mouse-like cheepings of innumerable birds. Vast flocks of Goldcrests are concentrating themselves in one spot; whilst woodcocks, singly or in pairs, dart aimlessly about, and fieldfares fill the air with a flutter of wings. Goldcrests are the smallest and frailest of British migrants, and even now they face the wild North Sea and essay to cross — they know not why or where. An all-absorbing impulse leads them on, as it has led innumerable generations of Goldcrests.

From Norland wastes of pine and spruce and fir they come in countless flocks; laying up no store of food, with no husbanding of strength; nothing but a longing to reach that far-off—they know not what. The mists rise from the sea, Norwegian heights begin to don white caps, and insect food is fast disappearing beneath tunnelled bark for its long winter sleep. The tiny wings grow restless, and wait only now for the night. And this, perhaps, is the strangest thing of all : When darkness has fallen, when winds are high and contrary, when the waterways of the fjords are boiling with foam—then it is that these frail things launch themselves on the storm and into the night. "From the land of snow and sleet they seek a southern lea."

In a Norwegian barque we are tossing off the Dogger Bank — betwixt that and the Galloper Lightship. The crosstrees and companion-ladder are covered with wheatears, titlarks, redstarts, a single blue-throated warbler, and hundreds of goldcrests. Thousands of the last species are coming and going, and others are beating out their little lives against the beacon-light. Vast flocks go on all through the night until dawn, when only stragglers blindly follow the same lines.

Or it is night, and we are waiting for the flights in the Lighthouse tower, where we have been since afternoon. The season is the time of the heaviest migration, and we are right in the track of the migrants. For nights past numbers of birds have been arriving and departing, and, as the sea-weather is "thick,"

more are expected. The daughter of the Light-
house-keeper is trimming the lights. The old
man himself is busily engaged in filling in
schedules which are next year to form the
materials from which to compile a Report on
the Migration of Birds. He shows something
more than an intelligent interest in his sub-
jects—knowing most by name, and describing
the flight and call-notes of the species he
fails to recognize so accurately as to render
ultimate identification by competent naturalists
certain. It is now nearly one o'clock. A
strong east wind blows over the North Sea,
with fog and drizzling rain. For hours flocks
of larks, starlings, mountain-sparrows, titmice,
wrens, redbreasts, chaffinches, and plovers strike
the light, and hundreds have fallen. Thousands
of birds are flying round the lantern—their
white breasts, as they dart to and fro in the
light-circle, having the appearance of a heavy
fall of snow. This is continued hour after
hour. The majority are larks, starlings, and
thrushes. A thousand must now have struck
the light and gone over into the sea. The
keepers of the Lighthouses and lightships say
that it is only on dark nights, snow, or fog, that

these casualties occur. When the nights are light, or any stars are visible, the birds appear to give the lanterns a wide berth.

Speaking of the night of the 28th of October, 1885, Mr. Gätke says, " We have had a perfect storm of Goldcrests, poor little souls! perching on the ledges of the window-panes of the light-house, preening their feathers in the glare of the lamps. On the 29th, all the island * swarmed with them, filling the gardens and all over the cliff—hundreds of thousands. By 9 a.m. most of them had passed on again." As to the state in which the little travellers arrive, Mr. Cordeaux tells me that, on a morning after an extraordinary flight, he saw numbers of Goldcrests on the hedgerows and bushes in the open marsh district of the Humber, creeping up and down the reeds in the drains; and at his lonely marsh farmstead they were everywhere busily searching for insects in nook and corner, fold-yard fence, cattle-shed, and stacks.

Just about the time of the ruin of the year vast flocks of Woodcock alight on our shores, passing southwards from their breeding-grounds. Like the rest of the migrants, the Woodcocks

* Heligoland.

travel in the night, and usually strike our seaboard about daybreak. Upon their first arrival, many of them are in an exhausted condition, and lie just where they have pitched until darkness again sets in. At nightfall they pass on. If the birds experience a fair passage, they do not touch our eastern coast, but, keeping well within the upper air, first drop in our western woods, or even those of Ireland.

The passage of this species is, curiously enough, invariably preceded by flocks of tiny Goldcrests ; and so invariable is the rule, that the latter have come to be called " Woodcock-pilots." The males precede the females by a few days, the latter bringing with them the young that have been bred that year. It is a point worthy of notice, and one upon which much confusion exists, that these migrants are usually in the very best condition. Soon after their arrival they disperse themselves over the leaf-strewn woods, and individual birds are known to resort to the same spots for many successive years. They seek out the warmer parts of the wood, and in such secluded situations rest and sleep during the day. At dusk they issue forth, in their peculiar owl-like flight, to seek their feeding-grounds. Like many birds, they have well-defined routes, and at twilight may be seen flying along the rides and paths of the woods, or skirting the plantations.

Coppice-belts they love, especially such as contain spring runs. It is here that the birds most easily find food, the soft ground enabling them to probe quickly and to a depth in search of earthworms. These constitute their principal diet, and the quantity that a single bird can devour is enormous. Sportsmen know that Woodcocks are here to-day, gone to-morrow. Where they were in plenty yesterday, not one remains.

Ireland affords the best shooting. There fifty brace have been shot in one day. This feat was the result of a wager, and the bag was made by 2 p.m. with a single-barrel flint-lock. The 'Cock were shot in an old, moist wood; and it is in such spots on the mild west coast that the " woodsnipe " finds its favourite haunt. In England the birds affect coppice-woods—frequenting most those which are wet, and such as have rich deposits of dead and decaying leaves. Than this, none of our birds conforms better or more closely to its environment. The browns and duns and yellows of its back have all their counterparts in the leaves among which it lies. Its protection lacks in one thing, however, and that is its large dark eye. This is full, bright, and obtrusive. It is not often that a special provision of this kind is injurious to its owner ; but the lustre which beams from the woodcock's eye is apt to betray its presence, perhaps slightly to negative the advantage of its protective colouring.

An interesting little bird, which every year comes to this country from the north, is the Snow-bunting. It travels from within the Arctic Circle, and so variable is its plumage that naturalists almost despair of ever getting a characteristic description. Indeed, so great a puzzle did this little stranger offer, that for long it stood to the older naturalists as three distinct species. Of course, we know now that the mountain, tawny, and snow bunting are one, and this because the birds have been obtained in almost every possible stage of plumage. They breed upon the summits of the highest hills with the Ptarmigan, and, like that bird, regulate their plumage to the pre-

T. Hutchinson lith.

C. E. Lodge

West,Newman imp.

Snow Buntings.

vailing aspect of their haunts. In this they succeed admirably, and flourish accordingly.

The Thrush family is one of the most numerous among British birds, and in winter comes much under notice. No birds suffer sooner or more acutely from severe weather; and the first frosts bring them about our gardens and homesteads in seach of food. Many of them are instinctively wild birds; but when their feeding-grounds are buried in snow or hardened by frost, their tameness becomes painfully conspicuous. And this applies not only to those which are resident throughout the year, but also to those which come to winter with us from the dense forests of Northern Europe.

Among these are Redwings and Fieldfares, both of which arrive on our coasts in autumn in countless numbers. Later they are occasionally joined by rare Bohemian Chatterers with their beautiful wax-like wing appendages, or by a little band of Crossbills. Our resident winter thrushes are the Throstle, Orange-billed Blackbird, Missel-thrush or Storm-cock, Ring-ouzel, and Dipper. The rest of the family is made up of a number of rare and occasionally occurring forms. Among these are White's

Thrush, the beautiful Rock-thrush, and the Black-throated Thrush.

Outside, here in the north, the lands are deeply covered with snow, and the usual supplies of food are cut off. The bright winter berries of the hedgerows have vanished, and the thrushes are more destitute than other birds. The evergreens of the gardens are full of them. In their aimless flying to and fro they shake the feathered rain from the snow-plumed branches, and many are so emaciated that they cower with drooping wings beneath the thicker shrubs. There the Blackbird darts out, as though to show how cleanly cut is his trim figure against the snow. He is more hardy than his congeners, and is self-assertive and bold wherever food is to be had. Few winter notes are so characteristic as his metallic "clink! clink!" coming through the thin frosty air at sundown. Even hardier than the Blackbird is the Dipper. This interesting brook-bird seems to revel in icy-cold water. See how he dashes through the spray and into the white foam, soon to emerge to his green mossy stone. Presently he melts into the water like a bubble, then reappears; and, after trilling out a loud wren-like song, again dips, jerking his body ceaselessly. By a rapid vibratory motion of his wings he drives himself down through the water, and by the aid of his wide-spreading feet clings to and walks among the pebbles. These he rapidly turns over with his bill, searching for the larvæ of water-flies and gauzy-winged ephemeræ. He searches the brook carefully downwards— sometimes quite immersed, at other times with his back out, and again with the water barely touching his feet. He does not always work with the stream, for I have frequently seen him struggling against it, but even then retaining his position upon the bottom.

It is from the dark pine forests of Norway and Sweden that the immense flocks of Redwings come that strike our coasts in October. At first small bands are seen under the hedges or in the fields, searching for the lower forms of animal life. At all times the Redwing is less a fruit-eater than its congeners, though the first hard frost immediately drives it to the hawthorns—the general resort of thrushes during times of scarcity. The fact that the Redwing seems ill adapted constitutionally to bear prolonged severity did not escape the keen eye of

Gilbert White, and he it was who first pointed out that it was among the first birds to suffer in winter. It is not unfrequently found so over-come by cold as to be quite unable to get away. The bird has its prettily descriptive name from the fact of its sides and lower wing coverts being light red or chestnut. For weeks past the laurels and holly-bushes of our garden have sheltered small flocks of these little northern thrushes—for the birds are gregarious, though never going in very great numbers. Even though unseen, the approach of the birds may be easily detected by the soft piping sounds which they utter in coming to roost at evening. Norwegian peasants call this bird the nightin-gale, from the deliciously soft notes of its song —thrush-like in their cadence, though resem-bling in mellowness those of the woodlark. The Redwing usually returns to its northern breeding-haunts about the beginning of April, though in cold springs it lingers long, and has even been known to breed in Britain.

The Fieldfare is another winter visitant, and constitutes the chief game-bird of the young gunner. This fine thrush does not make its appearance until a month later than the Redwing;

and upon its arrival we first note the "blue-jack" in upland pastures, where, if the weather is open, it finds a sufficiency of food in the form of worms and slugs. The more elevated tracts are agreeable to its habitual shyness; but upon the first coming of frost it descends to the lower grounds, and feeds upon the wild fruit-supply of the hedgerows. Large numbers of birds frequently roost together in some favourite spot—larch plantations, with thick undergrowth of coarse herbage being often selected. A bad habit to which the Fieldfare is driven during times of severity is that of drilling holes in the bulbs of turnips; and this not, as in the case of the wood-pigeon—when the root has been injured by insects or the bite of hares or rabbits.

Unlike the Redwing and Fieldfare, the Missel-thrush is resident; and, as a bird well known to dwellers in the country, has a host of provincial names. The "storm-cock" braves the severity of our hardest winters, and, like the rest of the thrushes, is a confirmed fruit-feeder. It feeds upon hips and haws; the berries of the ivy, holly, and yew; and upon those of the mistletoe, where this parasitic plant is found.

It is this trait that gives it its name. A break in the frost immediately sends it to the moist meadows, where it procures worms, snails, insects, and larvæ. Our earliest songster as well as earliest breeder, its loud song may often be heard from an ash-top during the most inclement January weather. Although shy and retiring, it becomes bold as the nesting season approaches, draws about the homesteads — frequenting orchards—and prefers sycamores and ashes at no great distance from dwellings. The characteristic call is a harsh " churr," hence the origin of " churr-cock."

The Throstle, the Mavis—" best beloved and most beautiful of thrushes ! " says Christopher North, watching one of these birds from his study-window at Elleray.

Looking out now upon the snow, how the bird with mottled breast seems oppressed with the sad labour of living ! What a different picture this from the warm summer evening, when a flood of song burst from every copse, and the Throstle was loudest and clearest of all ! Gone, too, are the shelled-snails, and here everything is iron-bound. Pugnacious no longer, the Blackbird, with satiny coat and orange bill, hops from beneath the laurels. His strong flight is laboured, his eye askance detects no food. Although an omnivorous feeder, competition is keen, and the Sparrows and Finches leave him but little. He pecks the hard ground; and the great red sun goes down without his parting "clink, clink!" Like the rest of the Thrushes, he is slowly starving.

One of the most beautiful and sprightly of British birds is the Goldfinch. . . .

A neglected field that has run to seed, covered over with nodding thistles and "horse-knops;" in its corners are bunches of groundsel and dandelion and plantain. It is rarely visited, and never stocked. Now and then a lad comes with a sickle and lays low the glowing pride of foxglove and thistle; but somehow he never works systematically, and hence never changes the aspect of that particular field. The characteristic flora of the spot still holds its own, and the weed-harvest of each year is greater than that of the last. Bunches of nettles and docks and campions hide the nesting-place of the ground birds, and under that rotten stump resides a colony of Hedgehogs. A pair of Larks have their nest under an overhanging tuft, and

Q

a Spotted Flycatcher seems ever to sit on a spray over the stream which runs down the bottom of the field. In our "intack" a pair of Corncrakes have taken up their abode, and give out their "crake, crake," far into the night. As we stand in the tall wet grass, the call seems to come from the middle of the field, then far out yonder; anon the bird runs nearly to our feet. We always love to hear the call, however, as it betokens summer, evening fishing, and long night rambles.

From the down and newly thrown earth the rabbits must have begun to breed; soon we shall see the young ones skipping about the mouth of the burrow, and pricking up their pinky ears when we endeavour to get a closer view. There has been a heavy shower of rain, and we meet a Hedgehog trotting off through the long grass. She just stops, turns up her coal-black eye, and reassuringly jogs on, knowing from past experience we are harmless. A pair of Partridges have made tracks through the grass, and probably have their oak-leaf nest under yonder clump of gorse. Here, too, the Meadow-pipits and Grasschats build, and life is everywhere. This is the aspect of our field in summer, when the hot breath meets one everywhere, and every tree displays masses of golden-green foliage. . . .

Six months have gone quickly by. The snow has fallen thickly for many days, and the pathways across the expanse are no longer to be seen. We wade wearily to our field, and stand by the wall. It is cold and forbidding, and we hardly care to enter. Life has forsaken it, and only over the white surface appear the dead crackling sprays of a few tall plants that dare to brook the blast. "Tweet, tweet!" comes through the cold thin air, almost startling amid the surrounding stillness. A flock of Linnets and Goldfinches!

And this is our second picture—A tall, nodding thistle-head, its once dark-green leaves shrivelled up and turned to grey, its purple flower-rays to russet-brown. Yet they contain ripened seeds. A Goldfinch hangs to the under surface, and a russet-breasted Linnet clings to the topmost spray. The two frail things are not unlike in form, though the Goldfinch is by far the handsomer bird. His prettily shaped beak is flesh coloured, as are also his legs. His head has patches of scarlet, white, and black, each well defined and setting off the other. The breast and back are of varying tints of warm russet-brown, the feathers of the wings picked out with white. His tail is alternately elevated and depressed as he changes his position, and the glowing patches of golden yellow on his wings are well brought out as he flits from spray to spray. Thus do the Linnet and Goldfinch go through the winter, together ranging the fields and feeding upon the seeds which they can pick up. Spring comes, and they separate, the one keeping to the woods, the other flying off to nest on the moorlands among the golden floods of glowing gorse.

The naked trees stand like ghosts against a cold grey sky; those pines seem hoary with their white weight of snow; the woods are painful in their very stillness, and the stillness is made more intense by the crackling sound of the snow as it is shivered from above by a courageous Squirrel that has ventured out to have a look at Nature in her wintry garb. Her pulse seems frozen, and outside all is cold, silent, and cheerless. Over the brown and bare woodlands, over the frozen streams and the hedges, descend the flakes of snow—soft, silent, and slow.

The Poacher will have a glorious time with his " gins " and " springes " and nets. Now he closely scans the weather, and will at evening pass under the wood and down by the " Hag " path. Heavily does he wade through the snow, his old black bitch doggedly following at his heels.

For hours from my look-out I have been sweeping with my glass the snow-plumed pines in search of a flock of interesting birds that do not appear. But in such weather as this the Crossbills always arrive. In severe winters I have never looked in vain for them in the pine wood. There they are! now on the upper, now on the lower branches; so tame that we may approach unheeded. The birds give out a constant twitter, and ever repeat their not unmusical call-notes. Never still, they are constantly changing position, fluttering from branch to branch, constantly sending down showers of cones and scales, and themselves hanging in every conceivable position. Nimbly they go, parrot-like, along the under sides of the boughs, climbing and holding with bill and

feet. What a babble of self-satisfied chattering comes from the feeding flock! What
wonderful adaption of means to an end in those
crossed mandibles! Every third cone or so
comes to the ground, but none are followed.
When one is secured it is held with the foot
upon the centre of a bough, and the bill quickly
invades the hard material. The birds feed for
an hour now, and return again late in the afternoon.

The severity of the weather in no way
affects them. Together they roam the fir
woods, feeding indiscriminately upon the cones
of fir, pine, and larch. Full of life and animation, their movements are ever changing. Their
plumage is various; bright red, orange, yellow,
and green are the coats of the individuals, but
no two seem quite alike. Once, and only once,
have they been observed on the confines of our
garden, and then feeding upon the scarlet fruit
of the Rowan or Mountain ash. Their partiality
to this food was amply testified by their completely denuding the trees. . . .

This morning we look upon a world unknown. The sun shines, and a rosy suffusion
lies over the landscape. All the fences are

buried deep, and the trees stand starkly outlined against the sky. Millions of snow-crystals
glint athwart the fields. Birds swarm in the
garden—the home birds more confiding, and
the wild birds tame. Tits hang to the suet
bags, and a general assembly flock to the corn-
sheaf. A Ring-ouzel flies wildly from a rowan-
tree, and four or five species of thrushes are
among the berries of the shrubs. So softly
winnowed is the falling snow, that it scarce
bends the few grasses and dead plants that
appear above its surface.

The kindly snow obliterates the torn and
abraded scars of nature, but it not the less
effectually reproduces the prints of her children.
To the light the snow reveals the doings of the
night. Does a mouse so much as cross, she
leaves her delicate tracery on the white coverlet.
Away from the homestead, rabbits have crossed
and recrossed the fields in a perfect maze. That
ill-defined " pad " tracks a Hare to the turnips.
Pheasants and Wood-pigeons have scratched for
mast beneath the beeches, and we find red blood-
drops by the fence. These are tracked to a
colony of Weasels in the old wall. Last night a
piteous squeal might have been heard from the

half-buried fence, and a little tragedy would be
played out upon the snow. Five Wild Swans
cleave the thin air afar up, and fly off with outstretched necks. The tiny brown Wren bids
defiance to the weather, darting in and out of
every hole and crevice, and usually reappearing
with the cocoon of some insect in its bill.

These delicate footprints reproduce the long
toes of the Lark, and those are the tracks of
Meadow-pipits. The hedge berries are mostly
gone; and the Redwing and Fieldfare have run
along the fence-bottoms in search of fallen fruit.
Those larger tracks by the sheep-troughs show
that the hungry Rooks have been scratching
near, and the chatter of Magpies comes from the
fir-tree tops. Scattered pine-cones betoken the
crossbills; and once in the fir-wood we caught a
glimpse of the scarlet appendages of the rare
Bohemian Waxwing. The gaudily coloured
Yellow-hammer shows well against the snow, and
bathes its orange plumage in the feathered rain.
How our British finches seem to enjoy frost and
snow! Certain it is that now their stores of
food become scant; but then they throw in
their lot with the sparrows of barn-door and
rick-yard. The bright bachelor-finch stands out

from his pure setting, and the Daws look black
against the snow.

Along the meadow brook a stately Heron has
left its imprints, the Waterhen's track is marked
through the reeds; and there upon the icy
margin are the blurred webs of Wild-Ducks.
A bright-red Squirrel runs along the white
wall; its warm fur showing sharply against
the fence. Naturalists say that the squirrel
hibernates through the winter; but this is
hardly so. A bright day, even though cold
and frosty, brings him out to visit some summer
store. The prints of the Squirrel are sharply
cut, his tail at times just brushing the snow.
The Mountain-linnets have come down to the
lowlands, and we flush a flock from an ill-
farmed field where weeds run rampant. When
alarmed the birds wheel aloft, uttering the
while soft twitterings, and then betake themselves to the trees. The seeds of brooklime,
flax, and knap-weed the Twite seems partial
to, and this wild-weed field is to them a very
paradise.

Just now, walking in the woods, the cry of the
Bullfinch is heard as perhaps the most melancholy
of all our birds; but its bright-scarlet breast

compensates for its want of cheeriness. A flock of diminutive Goldcrests rush past us; and in the fir wood we hear, but cannot see, a flock of Siskins. Higher up the valley, towards the hills, tracks of another kind begin to appear. On the fells we come across a dead Herdwick, trampled about with innumerable feet. We examine these closely, and find that they are of two species—the Raven and Buzzard. Further in the scrub we track a Pine Marten to its lair in the rocks. The dogs drive it from its stronghold; and, being arboreal in its habits, it immediately makes up the nearest pine-trunk. Its rich brown fur and orange throat make it one of the most lithely beautiful of British animals. A pair of Stoats or Ermines, with their flecked coats just in the transition stage, have their haunts in the same wood. From the snow we see that last night they have threaded the aisles of the pines in search of food.

This clear-cut sharp track by the fence is that of the Fox. Later we see the beautiful buoyant creature bounding over the snow in graceful leaps. Fleet and wild as the wind, his speed and play of muscle are hidden by his long soft fur. An exquisitely formed creature, we doubt as we look on him whether he is not worthy of the good things of the covert to which he is stealing. The most beautiful winter picture of this wintry morning is the red Fox on the white snow.

CHAPTER XXI.

" GIP."

A Sketch in Evolution.

Gip has stood by me through many an hour of adversity, and has always proved true as steel; and this is why I love him.

There is no trace of base metal in his composition—nothing but fine gold. And yet I hardly know why this should be, for there is little of *ton* about him. " Sir Windem," our other dog, came to us with a pedigree as long as his silky tail, but he was sadly lacking in moral qualities. Gip, however, has no pedigree to speak of—cannot even get on the paternal track of his father; and yet he has a capital nose.

So far as we can make out, there was only one being in the world who ever cared for Gip, and that was his mother. She belonged to a

bachelor; and, like Bartle Massey's " Vixen,"
brought disgrace upon herself by bringing
Gip into the world. But if this was feminine
folly, Gip was none to blame, and we never
mention it.

When he was only a few months old he had
developed several remarkable characteristics.
He seemed to have some subtle means of ana-
lyzing character, and so great faith did we have
in his powers of discrimination, that whenever
he seemed disposed to withhold his confidence
we invariably withheld ours; and subsequent
events showed that he was right.

But, with all his active intelligence, he is very
much a creature of habit. For instance, just
before lying down, he will turn round and round
and scratch the floor, even when the latter is perfectly smooth and hard. This action, meaningless in itself, is probably an inherited trait from
Gip's remote ancestors when they lived in a
wild state, and is practised now by jackals
and other allied animals in the Zoological Gardens. The action is intended to trample down
the grass or scoop out a hollow before lying-
down for the night, and in this manner the
animals just indicated treat their straw.

Gip's kennel companion, Sir Windem, a Laverock Setter, is even more a dog of habit than Gip. And this is somewhat strange, as he is come of an aristocratic line through Grouse, Heather Ranger, Sir Percival, Lord Raglan, Count Howard, and the rest of them. Sir Windem's fine feathering in the field — his pointing, backing, and down-charging—is only an elaboration of what Gip does daily.

Gip is what the sporting ones call " death " on Cats. If a strange pussy presents herself on the garden fence, he watches her intently, invariably keeping one of his fore-legs lifted the while. Most dogs do this, and it is doubtless that they may be ready for the next cautious step when they see or hear what they consider their legitimate prey. But that the action is often meaningless may be deduced from the fact that Darwin has seen a dog at the foot of a high wall, listening attentively to a sound on the opposite side with one leg lifted, and when of course there could be no intention of making a cautious approach.

Seeing that Gip's supply of bones is constant, it might be thought that there was no reason to bury superfluous ones. Yet this he regularly does. An apparent practiser of thrift, his "rainy day" has never yet come, nor is it likely to. Although he buries bones with the greatest assiduity, he was never known to disinter one. Consequently his bone-burying can hardly be an economical quality. Indeed, I am constrained to think that it is as useless as many of his self-imposed tasks, and that if it implies anything, it is the evolutionary suggestion that after all Gip may have descended from long-gone doggy ancestors that were once wolves, and jackals, and fennecs.

And if this be so, it brings us face to face with the fact that Gip's pedigree may be as long and respectable even as that of the silky-coated Sir Windem himself. Although this is speculation, I have a notion, from something I recently saw, that there is some truth in the idea of Gip's unwritten pedigree. We were out shooting, and the dogs were of the party. Sir Windem was working quietly and gravely; Gip noisily, and bustling about. A herd of cows, which before had been quietly feeding, now made towards us, forming a queue as they came. As we progressed the cows followed, always keeping their heads to the canine intruders; nor did they desist until the latter disappeared through the fence.

Gip's ancestors, wild dogs or wolves, were doubtless inveterate foes of cow-kind; and, by reason of inherited instinct, the antipathy is kept up to the present day—vaguely perhaps, but it is none the less there. This slumbering and almost unconscious antipathy remains, and common headway is made against the intruder.

When Gip returns home after a hard day's work in the open, he generally curls himself up before the fire and goes to sleep. And often the more exciting incidents of the day are repeated in his dreams. First his muscles twitch, and it is evident that there is excitement within. Then he pricks his ears, and indulges in short, sharp "yaps," and barks. It is evident he has got on the track of a hare; and, as the scent becomes warmer, his body seems hardly able to contain his emotions. But the crisis comes when Gip sights the hare, and with a bark bounds to his feet. Then comes the delusion. We all laugh; and Gip cannot bear to be laughed at. He slinks away with tucked tail, the very picture of utter humiliation and self-abasement.

His happy hunting grounds, however, are
not the only things of wakefulness of which
he dreams. Sometimes in his sleep he gives
out a low guttural growl, the scruff of his neck
is raised, and we know that he is about to meet
Tartar. These "twa dogs" always assume a
hostile attitude when they meet, calling for our
prompt interference, merely for the friendly
purpose of preserving the peace. This special
antipathy to Tartar is somewhat surprising; for
Gip is kindly disposed in all his domestic rela-
tions, even to Tabitha. Tabitha may hiss and
curl up her back, but he only looks on with a
grave kind of contempt.

Like most dogs, Gip scratches himself by a
rapid movement of one of his hind-feet; and,
when his back is rubbed with a stick to produce
a like sensation, so strong is the habit that
he cannot help rapidly scratching the air or
the ground in a useless and ludicrous manner.
Sometimes he will show his delight by another
habitual movement, namely, licking the air as
if it were my hand.

He has a language of his own, and it is quite
easy for any one acquainted with him to interpret his barks. He has a peculiar bark in

R

answer to a friendly greeting from another dog
across the fields ; and he emits a peculiar piping
sound in his more plaintive moods.   Intruders
he salutes with a bark in which there is nothing
but defiance, quite different from that by which
he shows his pleasure.   When he meets Tartar,
his ears are drawn back and laid flat upon his
head.   Upon such occasions the dogs have evil
intentions towards each other, and instinctively
protect the more delicate parts by rendering
them inconspicuous.   A movement with a like
object is shown by the tail tucked between the
legs.

Gip has just come into the room where I write,
and an insignificant act shows how greatly he,
in common with most dogs, is dependent upon
the sense of smell.   As he entered the door he
threw up his nose, and, after momentarily sniff-
ing the air, made a rush to where I usually sit.
A dog in searching for his master in a crowd,
invariably sniffs at the legs of all with whom
he comes in contact, trusting to his sense of
smell, and hardly using his eyes until he has
discovered the object of his search.

Gip's paternal responsibility upon one occasion came out in a truly remarkable manner.

He was on terms of acquaintance with a black
collie bitch, and, at the busy time of sheep
gathering on the fells, she gave premature birth
to four puppies in a bracken bed many miles
from home. One of these she carried in her
mouth over the whole distance, and deposited
among the hay in the barn. And at the same
time, by some subtle means, she acquainted
Gip with the critical situation, and the two
trotted back to the hills. Nor did they desist
until every puppy was safely housed. Just as
sheep and deer have the means of communicating
the presence of good pasturage, so do dogs tell
their friends of carcases upon the fells.

Of all inherited instincts the love of dogs to
man is one of the most pronounced. And of
this statement Gip is a bright example. There
is a quiet gravity about him that is almost
human. His expressive movements constitute
silent speech ; and if one act fails to convey his
particular want, he tries another. That he
reasons upon cause and effect is beyond question, in proof of which a hundred examples
might be given.

His life has its sunshine and its shade; or,
in other words, its rats and his master's dis-

pleasure. On the whole, Gip is a good dog, without making much profession. If there is any after-reward, Gip will have it abundantly; for he is a sympathetic and faithful friend, and what dog can be more?

CHAPTER XXII.

CAVE-HUNTING.

THERE was one hole among the rocks which Gip could never resist visiting in our morning walks. Its entrance was hidden by weeds and brambles, and was far down at the base of a towering limestone escarpment. I had generally to wait a long time for Gip, and when he deigned to return he was in a most disreputable condition. His nose and feet, as well as his shaggy coat, were covered with red loam, and at last I bethought me to see what so persistently attracted him.

Brushing away the weeds, I saw that Gip's hole was really a considerable fissure which extended far back into the limestone. In scratching the floor of the crevice he had disinterred several bones, the appearance of which excavations had been conducted in a quiet corner of the cavern, where only the finest film of carbonate had collected. With the point of a sharpened stick I broke away part of the incrustation, and revealed a band of clay containing more bones. Beneath this came a layer of red loam, richer in remains than the clay, if it did not wholly consist of organic matter. We had seen what we had seen; and Gip and I agreed that the matter should remain a profound secret until another day.

I am sorry to say that my generally faithful friend upon this occasion broke confidence; for that evening I saw him, covered with red loam, returning from the hole with another dog—not a scientific dog, but a foolish-looking poodle—and I afterwards found that the two had been prospecting on their own account.

Next morning, however, we went to our "Fairy Hole," this time provided with a short-handled pick, a spade, a pail of water, a brush, and a wooden box. Then we went to work. Gip vigorously, I sedately. We first removed several small blocks of limestone; and, after clearing the way, began to work the clay. This yielded an enormous quantity of bones,

but none of any importance. They consisted of
the remains of animals at present existing, together with those of Red Deer and Roe. It was
evident, then, that this deposit was a modern
one. We next set ourselves to clear away
completely the band of clay, and to examine
more closely the red loam beneath. This seemed
to consist almost wholly of ground bone, and
a strong lens afterwards proved that this was
so. Larger bones were plentifully embedded,
these sometimes occurring in layers. The
tusk of a Wild Boar was turned up, then the
humerus of some huge *Bos.* Horn cores were
frequent, bits of antler of the Red deer (indica-
ting animals of great size), and short " snags "
and " prongs " of the Roe. The remains of
the Badger and Wild Cat were common, also
those of several animals of the weasel-kind ;
and about three feet below the loam occurred
the best find of all. This we could not identify
at the time, but it afterwards turned out to be
the skull of a Bear.

As we gained in depth, tusks and the hardened
bone cases of wild boars became more frequent,
as also the remains of Wolves, a Beaver, and the
immense bones of Wild White Cattle. Subse-

struck me as unusual in outline. One of these was an elongated lower jaw, still holding a large canine tooth, which, after careful examination, turned out to be that of a Wolf.

Upon this little discovery Gip and I crawled along a tortuous passage until we found ourselves in a not inconsiderable cavern. For a moment this fairy hole reminded me of St. Brandan's Isle, in Kingsley's *Water Babies.* Trickling from the rocks far up in the dark recess, the water-bearing carbonate had spread a silver filament over all. Miniature stalactites of strange and fantastic forms depended from the roof, and bright bosses of the same glittering substance rose from the floor. At the entrance, where the light rushed in, it could be seen that fungi and golden mosses draped the dripping walls; the green of the moss was intensely green, the lichen tracery ravishing. There was a mass of Golden Saxifrage, the light shone through frail fern fronds, and a constant "drip" from above only served to make the silence more intense.

It seemed almost an act of vandalism to break through the stalagmitic floor; but after the wolf's jaw I was curious to follow up the clue. Gip's quently, from the *débris* which was thrown out, there were picked up some human remains, a bone needle, an awl, some beads, and an ear-ring.

Of course, the question at once suggested itself how these remains came to be here, and at the same time arose a difficulty. They were not those of quiet, peaceful creatures, but many were of the larger carnivora; and here were their bones strewed among those of animals upon which they were wont to feed.

The solution of this may probably be found in the fact that almost all carnivorous animals drag their prey to some quiet lair, there to devour it at leisure. Hence the accumulation of so many of the remains of deer and domestic animals. The same fact, of course, would apply to the human bones; also of the ornaments and imple-ments which were not unfrequently discovered. Then to account for the presence of the carni-vora themselves—it is well known that when these are wounded or overtaken by age they instinctively seek out some lair in which to die ; hence the curious compound of species in our fissure.

It may be, of course, that the rude hunters

of the times occupied these caverns, and thus
the bones had been conveyed there — the
domestic animals serving as food, while the
rest were the produce of the chase. And
this is to some extent borne out by the fact
that occasionally the palmate part of an antler
has afforded material for rude attempts at art,
which we find there depicted. These bone
sketches generally represent hunting scenes.

Our researches have also revealed flints,
arrow-heads, and chipped bones, but to un-
scientific workers like Gip and me these have
little meaning. We have succeeded, however,
in bringing to light the remains of an extinct fauna altogether different from that which
exists about us. We know that the Wild White
Cattle of the period grazed the green hillsides
of the valleys, and that bears of two species
kept the rocky fastnesses of the mountains.
Red Deer roamed the fell and corrie, while
the Roe in family parties trooped through the
old woods. The Beaver constructed its dam on
the banks of the secluded streams, and Wild
Boars wallowed by the margins of the reedy
meres. Badgers and Wild Cats were common,
and the Wolf a roaming marauder. Belts of

birch and oak clothed the hills, while most of the valleys were tangled swamps. Finally, the country was one vast forest and fell, and the few human dwellers of the time must have been primitive hunters.

CHAPTER XXIII.

HARES.

CONSIDERING how long the Hare has been known, there has been more unnatural history written concerning it than any other British animal.

It is said to produce two young ones at a birth; but observant sportsmen know that from three to five leverets are not unfrequently found. Even by some writers in what are called "standard" works it is stated that the Hare breeds twice or at most thrice a year. Any one, however, who has daily observed the habits of hares, knows that there are but few months in which leverets are not born. In mild winters young Hares have been found in January and February, whilst by March they have become common. They may be seen right on through summer and autumn, and even in October leverets apparently about a month old are not at all unfrequent. Of course exceptional seasons may account for this in some measure, but the same set of facts applies to ordinary years. Does shot in October are sometimes found to be giving milk; and in November old hares are frequently noticed in the same patch of cover. These facts would seem to point to the conclusion that the Hare propagates its species almost the whole year round—a startling piece of information to the older naturalists. Add to this they pair when a little over a year old; that gestation lasts only thirty days; and it may be seen how prolific an animal the hare may be. The young are born covered with fur, with their eyes open, and after about a month leave their mother and seek their own subsistence.

The Hare would certainly become abundant were it not beset by so many enemies. The balance could always be kept adjusted prior to the legislation of 1880; since which date, however, Hares have had no protection whatever. A shy and timid animal, it is worried through every month of the year. It does not burrow, and has not the natural protection of the rabbit. Although the colour of its fur allows it to conform in a marvellous way to the dead grass and herbage among whilst it lies, yet it starts from its " form" at the approach of danger, and from its size offers an easy mark. It is not unfrequently " chopped" by sheep dogs ; and in certain months hundreds of leverets perish in this way. They are often destroyed wholesale during the mowing of grass and reaping the wheat. For a short time in summer leverets especially seek this kind of cover, and farmers or their labourers kill great numbers with dog and gun ; and this at a time when they are quite unfit for food.

In addition to these causes of scarcity, there are others known to sportsmen, who have the remedy in their own hands. When harriers hunt late in the season, as they invariably do nowadays, many leverets are sacrificed without affording the least sport. Some are " chopped" in their forms ; and for every Hare that goes away probably three are killed in the manner indicated. At least that is the teaching of one who has had pretty wide experience. When hunting continues through March, master and huntsman assert that this havoc is necessary, in order to kill off superabundant jack-hares and so preserve the balance of stock. Doubtless there was reason in this argument before the present scarcity; though now it hardly holds good.

March, too, is a general breeding month, and hunting does in young entails the grossest cruelty. Coursing is confined within no fixed limits, and is sometimes prolonged unusually late. With some modifications, what has been said of hunting applies to coursing; and these things sportsmen can remedy if they please.

There is probably more unwritten law in connection with field-sports than any other pastime; and it obviously might be added to with advantage. If something is not done, the Hare will assuredly become extinct before very long. To prevent this, a " close time " is, in the opinion of most sportsmen, absolutely necessary. And the dates between which the animal would be best protected are probably the first of March and the first of August.

Poaching is almost, if not quite, as prevalent now as it has ever been; and the recent relaxation of the law has done something to encourage it. Poachers find pretexts for being on and about land which before were of no avail, and to the " moucher" accurate observation by day is essential to success. This is especially true in the case of hares and partridges. Each is local in its haunts and habits, and needs only to be closely watched to be easily captured.

As a rule, the village poacher knows the whereabouts of every Hare in the parish ; not only the field in which it lies, but the very clump of herbage in which is its form. In speaking of the poacher who makes hares his speciality, it is necessary to speak of his constant companion—the lurcher. Lurchers are pure crosses between Greyhound and Sheep-dog. The produce from these have the speed of the one, and the " nose " and intelligence of the other. Such dogs never bark, and, being rough-coated, are able to stand the exposure of cold nights. They take long to train, but when perfected are invaluable. The most successful poacher is the one who makes the greatest number of mental notes. In his walks abroad he watches hares feeding or at play, taking in their every twist and double. He examines all gaps, gates, and "smoots" through which they pass ;

and, so that he may leave no scent of hand nor print of foot, always approaches these laterally.

The Poacher looks out on the darkening night from his cottage, and when the time is favourable starts, striking right across the land. Frequented roads or by-paths he avoids. In a likely hedgerow he sets a couple of snares, not more than a yard apart; and if the field to be ranged holds a hare, he knows that it will take one of them. A dog is then sent to range the field, and the poacher has arranged that the wind shall blow from the dog across the hare's seat. This at once alarms her, and she comes lopping towards the fence. Something must be added to her speed, however, to make the wire effective, and this the dog supplies by closing in. Behind his snares the moucher, with hands on knees and still as death, waits for her coming. There is a rustle in the leaves, a faint squeal ; and the wire has tightened round her throat.

At gaps and gates a wide net is substituted for the snare, and often proves an engine of more wholesale destruction. These are the two common methods employed by the poacher.

s

His harvest is usually greatest in February and March, when hares are found in company.

Keepers and others interested in the preservation of hares ought to remember that a hare once netted can never be retaken in a similar manner. The moral of this is, that where poachers are troublesome, every hare on an estate ought to be taken in the manner indicated, and then turned loose.

CHAPTER XXIV.

THE only remaining herds of Wild White Cattle at present existing in Britain are those at Chartley, Chillingham, Cadzow, Somerford, Bickling, and Viangœ.

These constitute the last links of a once gigantic fauna, and probably owe their survival to the fact of their having been enclosed in parks and chases. These wild cattle are probably descended from the gigantic Urus—a huge *bos*, which once roamed through British forests and existed down to historic time. But, from whatever source descended, they abounded in the twelfth century in great numbers, occupying the forests which came close up to the gates of London. They were contemporaneous with Bears, Wolves, Wild Boars, and vast herds of Red and Fallow Deer.

At this distance of time it is difficult to realize the physical conformation of the country when there roamed over it the animals just enumerated.

The forests flanked the mountain ranges, stretching sometimes to their summits. Many of these were impenetrable; and especially does this refer to the tangled woods of the valleys. The chases, the forests of which they formed part, the swamps and fenlands— these occupied more than three-fourths of the country. As invasion set in from the south, the *feræ naturæ* retired northwards, seeking the wildest and densest security. And although the wild cattle for a time found shelter in the dark recesses of the great Caledonian forest, this was only for a time.

As they saw the native fauna fast disappearing, those to whom the king had granted tracts of forest bestirred themselves to preserve its mightiest members. These were the forest bulls (*Tauri sylvestres*, it is written), which were driven into paled or walled parks; and in this way many a country 'Squire preserved to his descendants direct produce of this once wild race. This form of preservation is best illustrated by the number of deer and deer-parks in Britain, though with this difference—that while the deer have survived, the so-called " wild " cattle have almost died out.

It would have been easy, in the case of the former, to prevent degeneracy by introducing animals from the Continent; and by the infusion of new blood to keep up the stamina of the indigenous stock. But no such recourse was possible in the case of the wild cattle, and consequently the remnant has gradually deteriorated, until now it has nearly ceased to exist. Through interbreeding, those at Chillingham have " a fine-drawn, almost washed-out appearance ; " and it has now become conspicuous that more males than females are born. But while this unhealthy sign is to be deplored, it to some extent offers a solution of the difficulty. It is proposed that the bulls from the few remaining herds be drafted out, and that an attempt be made to build up a new one in the park of some landowner who is sufficiently interested to undertake the experiment. If this prove successful, the result will be to preserve the type, and to give it added stamina and robustness.

In appearance the Wild White Cattle are
now somewhat smaller than domestic breeds, the
former having decreased in almost exact ratio
to the increase of the latter. They are covered
with long, shaggy hair, and have the hoofs,
muzzles, and ears black. In certain herds,
however (notably that at Chillingham), it has
been noticed that the extremities are red—
probably another sign of degeneracy. That at
Burton Constable, in Yorkshire, showed this
characteristic, though a studious endeavour was
made to preserve the black points. The cattle
here were somewhat larger than the more
known ones at Chillingham; and, probably,
owing to the richness of their Holderness pasturage, sometimes attained to sixty stones in
weight. A local characteristic also attached to
another Yorkshire herd, that at Gisbourne
Park. These were white, except the ears and
muzzles, which were brown or red. They
were without horns, strong-boned, and of low
stature.

A legend attached to this herd, that it had
originally been drawn from Whalley Abbey
by the power of music, much in the same
way that a number of Red Deer are said

to have been brought from the liberty of
the primitive forest to Hampton Court. The
members of this herd were exceedingly mis-
chievous, especially when guarding their young,
and approached any object of their resentment
in a very insidious manner. Although they
had bred with tame cattle, this herd also became extinct. They had become so delicate
from "in-and-in" breeding, that they had to be
housed in winter, and were quite tame. The
last cow and calf joined the herd of Mr. Leigh
at Lyme Park; and in 1859 the last bull of the
race was killed.

This last offers a typical example of the way
in which these semi-domesticated herds have
followed each other to extinction; though it is
interesting to learn that in some of the remaining herds the old characteristics of the original
race are still prominent. These have heads
slightly larger in proportion than those of
ordinary cattle, with large broad feet, and stout
legs.

It will not be surprising to learn that every
list compiled contains the names of fewer
herds than the last. For instance, there was
one at Lyme Park as recently as 1883, but

it had disappeared prior to Canon Tristram's
report. We look in vain for the name of
the herd at Kilmory House, Argyleshire,
which was present in the last list; and the
following have now, for a longer or shorter
period, ceased to exist: The herds at Ardrossan
Castle, and Auchencruive, Ayrshire; Barnard
Castle, Durham; Bishop Auckland, Durham;
Blair Athole, Perthshire; Drumlanrig Castle,
Dumfriesshire; Ewelme Park, Oxfordshire;
Hoghton Tower, Lancashire; Holdenby Park,
Northamptonshire; Leigh Court, Somerset-
shire; Middleton Park, Lancashire; Naworth
Castle, Cumberland; Whalley Abbey, Lanca-
shire; and Wollaton Park, Nottinghamshire.

CHAPTER XXV.

RECENTLY a large "school" of Whales, number-ing about four hundred, was seen disporting in the Bay of Firth, a few miles from Kirkwall. A number of fishermen hastily manned the boats, and commenced hostile demonstrations against the herd. Owing, however, to the roughness of the sea and the darkness of the night the luck of the fisher-folk was but small. The Whales were of the Caing, or Bottle-nosed species. It is twelve years since a herd visited the same shores.

This interesting occurrence serves to remind us that the Whales of the British seas are much more abundant than is generally supposed, not only in number, but in variety of species. Our knowledge concerning them is quite dispropor-tionate to the importance of the subject, as from

an economical standpoint the natural order
Cetacea is the most important which haunts the
seas. Owing to the element in which they
live, to migration, and to their invariably
keeping to deep water, the life-history of the
great Cetaceans is difficult to follow. Those
engaged in the Whaling industry of the northern
seas, and travellers in the same quarters, have
taught us much of certain species, and very
little of others. The first are those which are
most abundant or yield the greatest amount of
blubber; the second, the rarer forms, whose lives
are led far out in the deep waters, or such as
make but little return for chase or capture.

There are but few creatures which show
more admirably their adaption to environment.
Fitted as it is for an essentially acquatic existence, and spending its whole time in the water,
it is not surprising that the ancients invariably classed whales with fishes. The old naturalists, however, were well aware that the
young were brought forth alive, like those of
the rest of the mammalia, and that they were
nourished by the usual mammary organs.

The main physical features of the whale are its
distorted jaws, with upward-directed nostrils,

great bulk, and rudimentary limbs. The huge bulk of the creature is propelled by the flexible caudal fin, and whilst the body is rigid in front, it exhibits great mobility behind. The blow-holes are placed on the top of the head, and the animal can only respire when these are above water.

The larger Whales travel at the rate of about four miles an hour; but when pursuing their prey, or goaded by pain, they rush through the water at a much greater speed. They are aided in this by the broad and powerful tail, their chief organ of locomotion. Instead of being vertical as in fishes, this is horizontal, and the larger species can command immense driving power. The tail is also used as an offensive and defensive weapon. The smooth, shining skin is immediately underlaid by a thick coating of blubber—the great object of whalers. This is at once dense and elastic, and whilst it preserves the animal heat, it serves to reduce the mighty bulk of the whale and to bring it nearer to the specific gravity of the element in which it spends its existence.

An interesting trait in the economy of the Whale is the manner in which it suckles its young. In doing this, it partly turns on its side, and, the teats being protruded, sucking and breathing can proceed simultaneously.

Naturalists divide the Cetacea into two divisions, represented by the "whalebone" and "toothed" whales. In the former, the teeth are replaced by a series of great plates, and these, depending from the palate, constitute the *baleen* —the whalebone of commerce. The laminæ which comprise this, number about five hundred, are ranged about two-thirds of an inch apart, and have their interior edges covered with fringes of hair. Some of these attain to a length of fifteen feet. The cavity of a whale's mouth has been likened to that of an ordinary ship's cabin; and inside the surface conveys the idea of being covered with a thick fur. The soft, spongy tongue is often a monstrous mass ten feet broad and eighteen feet long. It might be thought that the whale, with its vast bulk, would want sea creatures of a large size to nourish it; but this is not so. Its chief food consists of minute molluscs—of Medusæ and Entomostraca—and with these its immense pasture-grounds in the Northern seas abound. In this connection will be seen the

beauty of the mouth-structure. "Opening
its huge mouth," says Professor Huxley, "and
allowing the sea-water, with its multitudinous
tenants, to fill the oral cavity, the whale shuts
the lower jaw upon the baleen plates, and,
straining out the water through them, swallows
the prey stranded upon its vast tongue."

One of the most important of the British
species is the Greenland Right-whale, a huge
creature, which amply rewards the Whaler with
a large supply of oil and whalebone, when he is
fortunate enough to fall in with it. It haunts
the cold water of the polar ice-fields, though,
in severe winters, it travels to the far south.
Although a rare visitant to European coasts,
whales of this species have occurred off Yar-
mouth and in the Tyne, whilst others are said
to have run aground on the Western Isles. On
the shores of Greenland it was once plentiful;
but here and at Spitzbergen it has been almost
exterminated by the Whalers. Baffin's Bay and
the neighbouring waters now constitute the
great whaling grounds, though, after summer-
ing here, the Whales move southward to winter
and produce their young. The breeding quarters are in the bays bordering Labrador; and

here the "schools" stay until the warm weather once more exerts its influence.

It is the chase of the Greenland Right-whale which has furnished so many stirring incidents from the Polar Seas. In its haunts, this species goes solitary or in pairs, and, as soon as one is seen, boats are lowered, and harpoons held in readiness. As opportunity offers, these are transfixed in the soft body of the huge creature, which at once dives. At every reappearance, harpoon after harpoon is thrown with deadly aim, until, from exhaustion and loss of blood, the creature finds itself incapable of further diving. The boats then draw near and attack it with lances, when the prey is almost assured. Great care has to be exercised, lest, in its dying convulsions—"flurry," as the Whalers call it—the maddened monster does not send boat and men, with a single swish of its mighty tail, flying like so many spars over the surface of the sea. If, however, the lances have done their work in the vulnerable parts, the monster turns over, and but little time is lost in getting to work on the carcase. First, this is hauled alongside; then the men set to work to "flense" it—to strip the

blubber, as well as the " baleen." Nothing now
remains but to remove the lower jaws, which
are rich in oil, after which the carcase, or
" krang," is set adrift. A dead whale attracts
a whole army of birds, beasts, and fishes, which
keep up an angry contest until the carcase is
demolished. Even the White Bear condescends
to be present upon these occasions, and revisits
the spot daily so long as the flesh is sweet.

The Whalers usually find their prey in the
" green water" belts of the ocean, and it is these
that furnish the huge Cetaceans with food. The
sea derives its peculiar tint from myriads of
tiny diatoms, which afford food to molluscs,
these again furnishing sustenance to the Whales.
The young are produced very early in spring,
though rarely more than one is observed following a female. This is suckled for a year, and,
as the " baby" whale is never far distant from
its mother, the Whaler always endeavours to
strike a " sucker." The two exhibit the warmest
affection for each other, and, if the latter be
struck, the mother allows herself to be har-
pooned rather than leave it. The colour of
this species is almost black—the young, bluish-
grey — the under parts of a creamy-white.

A large female whale of this species, killed in Davis Strait, measured along the curve sixty-five feet, of which the head was nearly a third; the greatest girth thirty feet. Specimens have been obtained weighing upwards of seventy tons, and such a creature may truly be said to be the largest which swims the ocean deserts.

The Atlantic Right-whale is another of the immense creatures of the Northern seas, though its range is more to the southward than the Greenland species. In fact, it is well known that at one time it was an object of pursuit round the British coasts; and the English Channel constituted a well-known fishery. Those who practised it here are said to have invented the harpoon, and, subsequently, to have taught the use of that weapon to Dutch whalers. Be that as it may, the chase of this species in our seas dates back to a very early period, and so rigorously was it prosecuted, that the animals were driven further and further north, and only found themselves free from molestation when they reached the polar seas. To prove that this is a different species from the last, it may be mentioned that old Icelandic manuscripts have two distinct names, one for a

Cetacean which appeared in winter, another for that seen only in summer. In proportion to the body, its head is much smaller than that of the Greenland Whale, and it is slightly less valuable to the Whaler. Another curious point of difference is that, whilst the more northern form is free from parasitic pests, the skin of the Atlantic species is invariably disfigured by cirripedes.

Dr. Hartwig states that barnacles often cover the whale in such masses that its black skin disappears under a whitish mantle; even seaweeds attach themselves to its jaws, floating like a beard, and reminding one of Birnam's wandering forests. The Whale has many noble enemies, but one of the vilest is a large sea-louse, which adheres to its back by thousands, and gnaws it so as to cover it frequently with one vast sore. In summer, when this plague is greatest, numbers of aquatic birds accompany it, and settle on its back as soon as it appears above the water to breathe.

Driven as it has been to Polar haunts, the visits of the Atlantic Right-whale to European or British coasts, are much more rare than formerly. When the southern whale fishery flourished, the Bay of Biscay was a great resort,'

T

and one of the latest authentic occurrences was
that of a female accompanied by a young one,
which entered the harbour of St. Sebastian.
The average length of this species is between
forty and fifty feet, and in colour it is almost
a uniform black.

The third of the British whales is the Hump-
backed, characterized by a dorsal fin and by
a number of longitudinal folds running along
the throat and belly. The head is comparatively
small, and the flippers extraordinarily large, the
latter in some cases measuring nearly a third of
the whole body. The Whalers call this species
the "Bermuda" Whale, and here, together
with its young, it may be seen from March to
May. Although not uncommon off the coasts
of Greenland in summer, not one is to be seen in
winter, as at this time it often wanders south.

Of the few which have actually occurred on
our home coasts, one at least is remarkable.
This was a female cast ashore near Newcastle,
measuring about twenty-six feet in length. The
contents of its stomach consisted of six Cormo-
rants, whilst a seventh stuck in its throat. Only
a few weeks before a fifty-eight feet Whale had
been washed ashore on Holy Island, but unfor-

tunately its species was not ascertained. A second specimen, of thirty-one feet, was taken in the estuary of the Dee, and its stomach was found to contain shrimps.

This species offers but little to the Whaler. Its blubber and whalebone are of inferior quality, and it is easily killed. The Greenlanders even attack it without harpoons, stealing along in their *kajaks*, and stab it with lances. Fish, molluscs, and crustaceans constitute its food in its native seas, and although it grows to sixty feet even, forty-five is perhaps the average length. Professor Hilljeborg says that often during calm weather it rests on the surface of the water, occasionally lying on its side, beating itself with its pectoral fins, as if trying to rub away something that annoyed it. Sometimes it jumps quite out of the water, turns round in the air, and falls on its back, beating itself with the pectorals. At times it appears quite fearless, and swims round about the boats quite near to them, as if they were its comrades. The young one follows its mother until she brings forth another, which is said not to take place every year, as very large young ones are sometimes seen with their mothers.

CHAPTER XXVI.

PIGEON-HOUSES AND FISH-STEWS.

ARCHÆOLOGICAL investigations in the North are constantly bringing to light remains of two institutions which once played a not unimportant part in the domestic economy of our ancestors. These are Pigeon-cotes and Fish-stews. They are mostly attached to old Manor-Houses and Baronial Halls, and probably at one time there were few of these strongholds without them.

To fully appreciate the value of their products we must go back to a time when the art of fattening cattle was but little understood and rarely practised. At this period the supply of animal food proved wholly inadequate to the demands of the community; for the stock fed out of doors in autumn was killed off by

Christmas, and but little fresh meat, except veal, appeared in the markets before the ensuing Midsummer. The more substantial yeomen and manufacturers provided against this inconvenience by curing a quantity of beef at Martinmas, the greater part of which they pickled in brine, the rest being dried or smoked by being hung in the chimney. Hogs were slaughtered after Christmas, the flesh being principally converted into bacon; and this, with dried beef and mutton, afforded a change to salt meat in spring. The fresh provisions of winter consisted of eggs, poultry, geese, and ill-fed veal, calves being conveyed to market when only a fortnight old.

These things constituted the food of the upper middle-class of the country districts, and it was only those still higher that could draw upon the resources of the " Culver-house " and Fish-stew. To them fresh fish and plump pigeons were always at hand to furnish a pleasant change from the hard salted meat. At this time the old British pastime of falconry had not yet gone out, and Duck, Heron, and Moor-fowl were often found at table. In the wilder parts of the North, Red Deer, Fallow, and Roe still held the woods or the hills, and venison in season was always welcome. Every religious house had its Fish-stew, as had the old Halls, and both monks and barons kept their "noble and deynteous fyssche" for fast-days, feasts, and general use. "Full many a fair partrich had they in mewe, and many a breme and many a luce in stewe."

The "partrich" was the Partridge, though it was much easier and profitable to keep domestic pigeons in store than wild game-birds in pound. There were good reasons other than luxury and comfort for setting such store by the delicacies of fresh fish and flesh. The prevalent diet has already been referred to, and there is no wonder that anything that would vary or palliate it was eagerly cultivated. But there was another reason. Those who were too poor to afford salt meat subsisted upon rye-bread and fish, and, what with the indigestible food of the rich and the too meagre diet of the poor, ague was of terrible frequency and leprosy common. These must be ascribed to the unwholesome food and privations of the people, for both disappeared as esculent vegetables came to be cultivated, and salted provisions fell into disrepute.

Macaulay reminds us of the Fish-ponds in which Carp and Tench were fattened for table; the Warrens of Conies, and the large round Dovecot rising in the immediate neighbourhood of the abodes of the great and wealthy, of the castle, the convent, and the manor-house.

To-day there is hardly a Hall or Keep in the North which does not show traces of its Dovecot or Pigeon-house; and when the actual cote is wanting, the name is almost sure to belong to some part of the demesne showing where it formerly stood. Thus there are dovecot—pigeon-house—and culver-house fields, upon which there are now no such buildings. Often the roost was situate among farm buildings, or sheltered by the massive masonry of the courtyard walls. The flocks of pigeons which this semi-domestication fostered often consisted of enormous numbers, and we are told that "corn were much destroyed by them." Hartlib, in his "Legacy of Husbandry," calculated that in his time there were twenty-six thousand Pigeon-houses in England. He further computed that by allowing a thousand birds to each Dovecot, and four bushels yearly to be consumed by each pair, the amount of corn

devoured by the birds annually would be
thirteen million bushels. If twenty-six million
birds existed at one time in this country the
number would be multiplied threefold in a
season. This army of pigeons would certainly
constitute an important item of diet for the
wealthier portion of the population, and the
encouragement and housing of the birds can
be well understood.

The great majority of the existing Cotes are
in a state of decay, though here is a description
of one still in preservation. The site is near
to where the outbuildings of the Manor-House
once stood, and is now surrounded by trees;
this would not be so, however, when it was
occupied, as the birds would not resort to a
Pigeon-cote in a wood. The Culver-house is
octagonal, and of dressed stone; the sides of
the octagon in the interior being upwards of
five feet. It has twelve rows of nests; the
lowest four feet from the floor; with a projecting slab in front as an effectual bar to
vermin or other destroyers. The remaining
rows have similar projecting ledges, though
narrower. The cells which constitute the nests
are nine inches in height, and L shaped. The

short limb or entrance is five inches by nine, and the long limb ten inches by five. About forty nests are in each row, making four hundred and fifty in all. The octagonal roof has a pent-house, with holes for the passage of the pigeons. Beneath are a potence and ladder; the ashlar work of the Pigeon-house being identical with that of the mansion near which it stands. The architect of the latter was Inigo Jones, and it was built during the last decades of the seventeenth century.

Pigeon-houses are exceedingly common in Persia, as in all parts of Palestine. In the latter country nearly every house has its Cote, those of the more wealthy being built with a number of earthen jars, which are roofed over. Each jar is supposed to accommodate a pair of pigeons, and many such exist in England. In the East the people who are too poor to possess Pigeon-cotes keep the birds in their houses, these entering by the doors and windows. Clouds of them may be seen over the fields, those of different owners flocking together. All the kinds of British wild pigeon are found here, as well as other species and Turtle-doves. One of the most common kinds is a bird nearly akin to our Rock-dove. The valleys of the limestone districts swarm with them, and they breed among the cliffs and escarpments in prodigious numbers. In short, Palestine is a country of pigeons, and travellers can fully appreciate the great number of references to these birds in the Scriptures.

The Monks knew a great deal about the cultivation of fish, their breeding and rearing, their subsequent management and fattening in the Stews. This art is still much practised in certain European countries, where the conditions to-day are like those which prevailed in England two or three centuries ago. Most of the fish fattened were used upon fast-days. Every Monastery in the country had its Stew, and such Manor-Houses as were occupied by Catholic families. As well as introducing many rare and dainty fish from the Continent, the monks reduced the cultivation of fish-ponds to a science. It was customary to have a series of these, which grew in turn fish and vegetables. The ponds were so arranged that they could be drained at will; and periodically the water was run from the first, the fish being caught as it emptied, and transported to the second. Number one was then planted with oats, barley, or rye-grass, the crop being reaped as it matured; as winter came round, it was re-stocked with fry and yearlings. By this process the pond was not only sweetened, but its supply of food was greatly improved; with the result that the fish turned into it grew and fattened in an extraordinary manner. When each had been worked in rotation, one was growing a crop of vegetables; another fry and yearlings; and the third breeders and fish fattening for market. Suitable weeds were grown about the margins of the ponds, and in many instances much care was taken in the matter of feeding. As the fish grew to a large size, they were netted and placed in the actual Stew. An ingenious contrivance for taking them out at pleasure was a strong wooden box, having holes in the bottom, and which was sunk where the water was deepest. As required, the box was wound up by a chain, contents and all.

A great variety of fish were kept in the ponds, and fatted in the Stews when these were in vogue. Among them were Carp, Tench, Pike, Eels, Trout, and many others. Thought

was given to the habits of these, and while
Tench and Eels succeeded best in mud, Carp were
kept on gravelly bottoms. Certain fish devoured
the spawn of others, and care had to be taken
to protect one species against its neighbours.
On this account, Carp and Tench thrive and
breed best where no other fish are put with them
into the same pond.

Walton reminds us that when re-stocking a
pond with Carp, it is necessary to put in three
milters for one spawner, and that the pond
should have certain characteristics. It should
be stony or sandy; warm and free from wind;
not deep, and have willows and grass on
its sides. Then he notes that Carp usually
breed in marl-pits, or such as have clean clay
bottoms, and are new. The Pike, or "luce" as
it was called, was in great request for fattening in Stews, as it grew with great rapidity.
The Char, one of the most beautiful and dainty
of British fishes, is said to have been introduced
by the monks, as doubtless were the various
species of Carp.

Carp-culture on the Continent is quite an
important industry, and in ancient days these
fish were in great repute for the table. Of late,

but little attention has been paid to its cultivation; but in the *Boke of St. Albans* it is described
as "a deynteous fyssche, but scarce." It is
little wonder that the Monks were alive to the
merits of Carp, for no fish was better adapted
to thrive in the Stews and fishponds where the
finny live-stock was usually kept.

In France and Germany, Carp culture is
quite an important industry, and a great many
persons are engaged in it — both men and
women. On certain days the fattening ponds
are emptied, and the Carp conveyed in carts
to the nearest market towns. It is well-known
that a Carp takes a good deal of killing;
and though being tightly packed in straw for
a whole day, and jolted downhill for hours,
may strike him as a novel experience, it does
not do him the least harm.

The Tench being a fish of a contented mind,
almost any kind of conditions suit his tem-
perament. As a store-fish he is invaluable,
and in any case gives but little trouble. Of
all fish of pond or Stew, the Tench is the most
accommodating. Like the Carp, he can be conveyed long distances to market, and if not sold
may be brought back to await another occasion

for sale.   Bream, as a stew-fish, has been
appreciated since the time of Chaucer ; and
Walton, in his admiration, refers to him as
"large and stately."   Bream, like Tench, are
fond of still, quiet waters, with soft soil bottoms,
in which they find their chief sustenance.
This fish has been known to attain to seventeen
pounds in weight, though this is somewhat
exceptional.   There is a French proverb to the
effect that " He that hath Breams in his pond
is able to bid his friend welcome ; " and if the
bream is toothsome, he is equally good as a
sporting fish.

These are some of the fresh-water fish which
once occupied the Stews in this country, and
might with profit do so again.

CHAPTER XXVII.

ALTHOUGH Britain can show no parallel either in number or brilliance to the living lights of the tropics, we are not without several interesting phosphorescent creatures of our own. Those whose business leads them abroad in the fields and woods through the short summer nights are often treated to quite remarkable luminous sights. One night we were lying on a towering limestone escarpment, waiting to intercept a gang of poachers. The darkness was dead and unrelieved; and a warm rain studded every grass-blade with moisture. When the day and sun broke, this would glow with a million brilliant prismatic colours, then suddenly vanish. But the illumination came sooner and in a different way. The rain ceased,

and hundreds of living lights lit up the sward.
In the intense darkness these shone with an
unusual brillancy, and lit up the almost impal-
pable moisture. Every foot of ground was
studded with its star-like gem, and these
twinkled and shone as the fire-flies stirred in
the grass. The sight was quite an un-English
one, and the soft green lights only paled at
the coming of day. One phase of this interesting phenomenon is that now we can have
a reproduction nightly. The fire-flies were
collected, turned down on the lawn, and their
hundred luminous lamps now shed a soft lustre
over all the green.

Why our British fire-flies are designated
" glow-worms " is difficult to understand.
Lampyris noctiluca has nothing worm-like about
it. It is a true insect. The popular misconcep-
tion has probably arisen in this wise. The
female glow-worm, the light-giver, is wingless ;
the male is winged. The latter, however, has
but little of the light-emitting power possessed
by the female. Only the light-givers are
collected, and being destitute of the first
attribute of an insect—wings, are set down
in popular parlance as worms. Old mossy

banks, damp hedge-rows, and shady woods are
the favourite haunts of the little fire-flies, and
the warm nights of the soft summer months
most induce them to shed their soft lustre.
Some widowed worm or fire-fly flirt may stud
her luminous self on the darkness even into
dying summer or autumn. But this is unusual.

It is not definitely known what purpose is
served by the emission of the soft green light,
but it has long been suspected that the lustre
was shed to attract the male. And this seems
reasonable. Gilbert White found the glow-
worms were attracted by the light of candles,
and that many of then came into his parlour.
Another naturalist captured by the same light
as many as forty male glow-worms in an evening. Still another suggestion is that the
phosphorescence serves as a protection or means
of defence to the creature possessing it, and an
incident which seems to support this view has
actually been witnessed. This was in the case of
a carabeus which was observed running round
and round a phosphorescent centipede, evidently
wishing, but not daring, to attack it. A third
explanation of the phenomenon is that it serves
to afford light for the creature to see by. A

somewhat curious confirmation of this is the fact
that in the insect genus to which our British
fire-flies belong—the Lampyridæ—the degree
of luminosity is in inverse proportion to the
development of the vision.

Fire-flies glow with greatest brilliancy at
midnight. Their luminosity is first seen soon
after dark. " The glow-worm shows the matin
to be near, and 'gins to pale his effectual fire."

As the insects rest on the grass and moss,
the difference in the amount of light emitted is
most marked. While the luminous spot indicated by a female is quite bright, the males
show only as the palest fire. When on the
wing the light of the latter is not seen at all.
Heavy rain, so long as it is warm, serves only
to increase the brightness. The seat of light in
the glow-worm is in the tail, and proceeds from
three luminous sacs in the last segments of the
abdomen. The male has only two of these sacs,
and the light proceeding from them is comparatively small. During favourable weather the
light glows steadily, but at other times it is not
constant. The fire-flies of the tropics—even
those comprising the genus Lampyridæ—vary
to the extent that while certain species control

their light, others are without this power. The light of our English glow-worm is undoubtedly under its control, as, upon handling the insect, it is immediately put out.

It would seem to take some little muscular effort to produce the luminosity, as one was observed to constantly move the last segment of the body so long as it continued to shine. The larva of the glow-worm is even capable of emitting light, but not to be compared to that of the developed insect. Both in its mature and immature forms the *Lampyris noctiluca* plays a useful part in the economy of nature. To the agriculturist and fruit-grower it is a special friend. Its diet consists almost wholly of small shelled snails, and it comes upon the scene just when these farm and garden pests are most troublesome.

British fire-flies have probably never yet figured as personal ornaments to set off female beauty. This is, and has long been, one of their uses to the dusky daughters of the tropics. They are studded about the coiled and braided hair, and perform somewhat the same office as the diamond for more civilized belles. Spanish ladies and those of the West Indies enclose fire-flies in bags of lace or gauze, and wear them amid their hair or disposed about their persons. The luminosity of our modest British insect is far outshone by several of its congeners. Some of these are used in various ways for illumination, and it is said that the brilliancy of the light is such that the smallest print can be read by that proceeding from the thoracic spots alone when a single insect is moved along the lines.

In the Spanish settlements, the fire-flies are frequently used in a curious way when travelling at night. The natives tie an insect to each great toe, and on fishing and hunting expeditions make torches of them by fastening several together. The same people have a summer festival, at which the garments of the young people are covered with fire-flies, and, being on fine horses similarly ornamented; the latter gallop through the dusk, the whole producing the effect of a large moving light.

Another phosphorescent little creature found commonly in Britain is a centipede with the expressive name *Geophilus electricus.* This is a tiny living light which shows its luminosity in a remarkable and interesting fashion. It may not uncommonly be seen on field and garden paths, and leaves a lovely train of phosphorescent fire as it goes. This silvery train glows in the track of the insect, sometimes extending to twenty inches in length. In addition to this, its phosphorescence is exhibited by a row of luminous spots on each side its body; and these spots of pale fire present quite a pretty sight when seen under favourable circumstances. It was stated that the light-giving quality of the fire-flies might be designed to serve them to see by; but this fails to apply to the little creature under notice, as it is without eyes.

There are still other British insects which have the repute of being phosphorescent; though these cases are not yet quite authenticated. Among them are the male-cricket and "daddy-long-legs," both of which are reported to have been seen in a phosphorescent condition. But if there is a dearth of phosphorescent land creatures which are native, this has no application to the numerous living lights of our Southern British seas. Among marine animals the phenomenon is more general and much more splendid than anything which can be seen on land, as witness the following picture by Professor Martin Duncan: "Great domes of pale gold with long streamers, move slowly along in endless succession; small silvery discs swim, now enlarging and now contracting, and here and there a green or bluish gleam marks the course of a tiny, but rapidly rising and sinking globe. Hour after hour the procession passes by, and the fishermen hauling in their nets from the midst drag out liquid lights, and the soft sea jellies, crushed and torn piecemeal, shine in every clinging particle. The night grows dark, the wind rises and is cold, and the tide changes; so does the luminosity of the sea. The pale spectres sink deeper and are lost to sight, but the increasing waves are tinged here and there with green and white, and often along a line, where the fresh water is mixing with the salt in an estuary, there is brightness so intense that boats and shores are visible. But if such sights are seen on the surface, what must not be the phosphorescence of the depths? Every sea-pen is glorious in its light; in fact, nearly every eight-armed Aleyoriarian is thus resplendent, and the social Pyrosoma, bulky and a free swimmer, glows like a bar of hot metal with a white and green radiance."

The Red Grouse of Britain is an indigenous game-bird, and is found nowhere else in a wild state ; though it is probably only the local representative of the white Willow-grouse of Northern Europe. It is somewhat strange, that a bird with such a limited range as the former should be almost wholly dependent upon one or two wiry shrubs for its sustenance. These are both mountain or moorland plants, generally known as heather; the one the common Ling (*Calluna*), the other the purple Heath (*Erica*).

Young Grouse just hatched find their choicest food in the extreme ends of ling and fine-leaved heather; and so great an influence has different varieties of food upon the birds as to change the colour of their plumage. Grouse which are found feeding upon old heather are almost black, whilst those with a supply of young purple shoots are much more brightly plumaged. So great is the variation, that many birds have peculiar local characteristics. Thus it is said that Grouse from the Hebrides and Wigtonshire are smaller and lighter than those from eastern moors ; Perthshire birds are smaller and darker than those of Argyllshire ; whilst

CHAPTER XXVIII.

HEATHER-BURNING.

DURING the past fifty years Scotch shooting rents have enormously increased in value; in some cases they are twenty times higher than they were half a century ago. This is owing to the fact that Grouse and Deer have increased in numbers just as remarkably. The change is to be accounted for by two things: The afforesting of the straths and glens, and the periodical burning of the heather.

Ground which twenty years ago fed five Grouse now supports fifty; and much the same may be said of Deer. Before the era of heather-burning and draining, the very names of some of the largest estates in the Highlands were unknown; and the game made no return whatever to the laird beyond the produce of his private shooting.

in Lanark, Renfrew, and the border counties
they are occasionally as light-coloured as partridges. Welsh birds are said to be large in
size and light in colour; those from the north
of England more rufous; whilst Irish birds are
lighter, with browner legs, and a yellowish-
red tinge in the plumage. These peculiarities
are undoubtedly due to food-supply; therefore nothing ought to be left undone to secure
the best and soundest quality of heather and
ling.

Undrained moors and a preponderance of old
heather are fatal. Wet cold land ought to be
made warm and dry, and the heather periodically burned. Grouse will not lie in tall rank
heath, neither do they care to nest in it; and
young Grouse that eat decayed fibres die of
indigestion. They love to nest in grass and
ling a few inches in height, and it is this kind
of ground-cover that produces the strongest
and healthiest coveys—a state of things which
follows immediately upon heather-burning. As
much discretion ought to be used in burning
in proper rotation as in sowing successive crops.
Where there is nothing but rank herbage, the
birds must not be deprived of all their cover

at once. That which is long and benty, bare and thin, ought to be burned first; this will be found to be loose and open below, the tough twigs furnishing neither food nor shelter. The best proportion to burn is about an eighth of the whole moor; and, after the annual rotations are once completed, it will be found that the shrub never gets so long as to be injurious to game.

The good effects of burning are manifold. The ash of the burnt plant aids in producing finer and sweeter herbage, and soon delicate heath-shoots peep through—the flowers of these sweet-smelling and brightly coloured. When this is done, the grouse-packs flock from the neighbouring moors, as do sheep and deer. Among growing ling and mountain shrubs, Grouse at certain seasons feed upon the seeds of coarse moorland grasses; also upon the leaves and berries of the black and red Whortle-berry, Crowberry, and occasionally oats if these are grown on the confines of the moor. The growth of berry-bearing bushes is encouraged by clearing, and, when once the experiment is progressing, heather of good quality and of every stage of growth will be the result.

Experience has taught that the yield of heather is by no means alike on different soils; and this fact is one that must always be taken into account. If heath of ten years' growth be allowed on land, harm must ensue. The birds will probably die of liver-disease.

The evidence of specialists as to the economy of heather-burning all leads to the same conclusion. In Rhidorrock Forest, in Ross-shire, before the era of clearing by fire, very few birds could be seen on the hills, and most of these were old cocks. But the year next after burning " hundreds and hundreds" came from all the neighbourhood around ; and afterwards increased amazingly. In another case, where at first it was complained that the heather had been over-burned, nearly twice as many birds were shot on the ground as had ever been shot before. As the season advanced, the birds came down in large packs to feed upon young shoots of heather springing on the burnt ground.

The wisdom of heather-burning is just as evident in the case of sheep and deer. Periodical burning supplies plants with the elements of nutrition upon which the rootlets immediately seize. The alkalies produced by combustion invigorate the ground and promote healthy plant-life ; whilst the roots are in nowise injured by burning. No ordinary fence will prevent grey-faced Scotch sheep from clearing it, if they are upon a grazing that has not been burned and adjoining one which has. Both sheep and deer have the means of communicating to each other the presence of good pasturage. It is in evidence before a Select Committee of the House of Commons that all the sheep in a neighbour-hood rushed to the burnt heather-patches ; the animals could not be kept to their own ground. It was also pointed out that if once two or three sheep find the better food, they bring others ; "they all rush to it."

Even in districts where it is an innovation, the economy of heather-burning is coming to be recognized, and the prejudice against it has now nearly died out. The benefits that follow are known, and both farmers and sportsmen are willing to recognize them ; so much so, in fact, that during a late period of drought, moors which were supposed to have been set on fire by sparks from passing engines are known to have been fired designedly by people who had before opposed burning and now had not moral courage enough to own their conversion. Of course much harm is done by indiscriminate clearing of this kind.

Ling ought to be burned in regular strips or patches, and to do this the flames have to be closely controlled. Seeing that advantage is mutual, keepers and farmers ought to work amicably together, so as to obtain the best results. In Scotland, an unwritten rule holds that the shooting tenant shall provide two assistants for every one of the farmer; or that the former shall pay the latter the expenses incurred in like proportion ; and this plan is generally found to work well. It may not be generally known that it is illegal to burn heather after the 26th of April, for reasons which at once suggest themselves. Now this "close time" sometimes seems to the sheep-farmer hard legislation indeed. In winter and early spring the mist-caps stick closely to the hill-tops and the ground remains saturated. Particularly is this so where wet "floes" abound; it is almost impossible to burn the heather on these patches. The present restrictions are probably wise ; but at the same time it is well known that ground cannot be burned in one year out of five on wet and marshy moorland; and it frequently happens that the last day or two of the season are the only ones in the whole year when burning is practicable. This leads to hasty work, which is often productive of harm and irritation.

THE END.

www.ingramcontent.com/pod-product-compliance
Lightning Source LLC
Chambersburg PA
CBHW021506210326
41599CB00012B/1150